149.6

Deer Antlers

REGENERATION,
FUNCTION, AND
EVOLUTION

Deer Antlers

REGENERATION, FUNCTION, AND EVOLUTION

RICHARD J. GOSS

Division of Biology and Medicine
Brown University
Providence, Rhode Island

Illustrated by Wendy Andrews

1983

ACADEMIC PRESS

A Subsidiary of Harcourt Brace Jovanovich, Publishers

New York   London
Paris   San Diego   San Francisco   São Paulo   Sydney   Tokyo   Toronto

ACADEMIC PRESS, INC.
111 Fifth Avenue, New York, New York 10003

United Kingdom Edition published by
ACADEMIC PRESS, INC. (LONDON) LTD.
24/28 Oval Road, London NW1 7DX

Library of Congress Cataloging Publication Data

Goss, Richard J., Date
    Deer Antlers: Regeneration, function, and evolution.

    Includes bibliographical references and index.
    1. Antlers. 2. Deer. I. Title.
QL942.G67 1983   599.73'57   82-22795
ISBN 0-12-293080-0

*To the memory of
Daniel Southwick*

Contents

Preface   **xiii**

1   Introduction

Text   *1*
General References   *5*

2   The Diversity of Deer

Fallow Deer   *8*
Red Deer   *10*
Sika Deer   *11*
Sambar   *13*
Barasingha, or Swamp Deer   *14*
Eld's Deer, or Thamin   *16*
Schomburgk's Deer   *17*
Rusa Deer   *19*
Hog Deer   *19*
Axis Deer, or Chital   *20*
Père David's Deer, or Mi-lu   *21*
Muntjac   *23*
Tufted Deer   *24*
Musk Deer   *25*
Chinese Water Deer   *26*
Roe Deer   *27*
Wapiti, or American Elk   *29*

Moose   31
Reindeer and Caribou   33
White-Tailed, Black-Tailed, and Mule Deer   35
Brockets   39
Pampas Deer   40
Marsh Deer   41
Huemul   41
Pudu   42
Cervid Classification   42
References   47

3   Horns and Tusks

How Horns Grow   57
The Improbable Pronghorn   62
The Histogenesis of Horns   66
Tusks   69
References   70

4   The Evolution of Deer

Fur-Covered Horns   73
Nondeciduous "Antlers"   77
The Earliest Antlers   84
The World's Largest Deer   87
Continent without Deer   92
Deer in the Americas   94
Recapitulation   96
References   98

5   The Social Significance of Antlers

Rutting Behavior   101
Phylogeny of Function   108
Why the Velvet Is Shed   111
Theories of Antler Casting   113
Ultimate Explanations   116
References   119

6   A Fawn's First Antlers

Pedicle Development   122
Antlerogenic Transformation   124

Fawns versus Yearlings   125
Photoperiod and Hormones   126
Histogenesis   127
References   131

7   Developmental Anatomy of Antlers

Antler Casting   137
Histogenesis   141
Wound Healing   147
The Nature of Velvet   148
Zones of Differentiation   153
Chondrification   154
Ossification   158
Systemic Effects of Antler Ossification   161
Vascularization   162
Innervation   166
References   168

8   Regeneration

Modes of Growth   172
The Spectrum of Regeneration   174
Mechanism of Blastema Production   177
Epimorphic Regeneration in Mammals   180
Strategy and Prospects   190
References   192

9   Abnormal Antlers

Genetic Effects   193
Systemic Influences   196
Injuries in Velvet   203
Pedicle Wounds   206
Do Nerves Affect Antler Development?   208
Nonexistent Mistakes   211
Horn Abnormalities   212
References   215

10   The Case of the Asymmetric Antlers

The Prevalence of Asymmetry   218
Is Asymmetry Genetic?   220

The Brow Tine Mystery   222
References   229

11   **Light and Latitude**

Reversed Seasons   232
Frequency Changes   234
The Role of Latitude   238
The Tropical Paradox   240
Circannual Rhythms   243
References   246

12   **Internal Influences**

Testosterone   249
Estrogen   254
Progesterone   255
Pituitary Hormones   256
Adrenal Steroids   257
Thyroid Hormone   258
The Effects of Pinealectomy   258
Nutrition versus Genetics   260
References   263

13   **Castration**

Methods and Degrees of Emasculation   268
The Castrate Antler   270
The Antleroma: A Model for Tumor Growth?   278
Antlers and Cancer   281
References   281

14   **Antlered Does and Antlerless Bucks**

Reindeer and Caribou   284
Incidence of Female Antlers   285
The Fertility of Antlered Does   287
Experimental Induction of Female Antlers   290
The Absence of Antlers in Males   293
References   295

15 Medicinal Uses of Antlers

Historical   298
The Antler Business   299
Harvesting and Preparation   300
Prescriptions   302
Scientific Investigations   303
Ethical Pros and Cons   304
References   305

Index   307

Preface

He who undertakes a book on deer antlers is inevitably asked how so narrow a subject could fill an entire volume, not to mention interest very many readers. Specialized subjects, however, have ramifications far beyond their own apparent limits, and implications sometimes of crucial importance to the solution of fundamental problems. Deer antlers are no exception.

How such improbable appendages could have arisen in the first place has been a challenging problem for evolutionary ecologists. The significance of antlers in the social lives of deer seems to generate theories in inverse proportion to available facts. As secondary sex characters, antlers are profoundly and singularly influenced by sex hormones. Their annual demise, loss, and replacement makes them the only mammalian appendages capable of complete regeneration. The mechanism by which these "bones of contention" grow and differentiate into such magnificent morphologies is a source of wonder and curiosity. Although antlers may not be the answer to cancer, it is perhaps significant that they grow more rapidly than tumors and give rise to cancerlike outgrowths following castration.

Despite their many exceptional attributes, antlers have always been a relatively neglected field of investigation. No more than a handful of scientists has been seriously involved in antler research at any one time in the past century, yet there has hardly been a year when deer antlers were not being studied by someone. Both professional biologists and laymen have contributed knowledge to this field. Unlike more conventional areas of scientific investigation, not all of the sources of information are to be found in technical journals and books. One's knowledge can be enhanced by the anecdotal accounts of people who work closely with deer. Zookeepers and hunters often turn out to be rich sources of insight, because in the need to understand the objects of their care or quest they must become astute observers of nature.

Perhaps one reason for the relative paucity of scientific investigations of antlers is that most biologists work on traditional laboratory or domesticated animals, or on wild ones which do not lend themselves to controlled observations on specific individuals. To study the growth of antlers in deer, one must manage to keep these wild animals in captivity, a formidable challenge to the enterprising zoologist. It is a rewarding challenge, however, because in such an unexplored field as this, almost anything one learns is a new discovery. A major source of frustration is that to repeat an observation on growing antlers one must wait a whole year for the next cycle. Yet the disadvantages of doing research on deer are far outweighed by the opportunities to become acquainted with these graceful and spirited animals.

My own love affair with deer dates back almost 25 years. It derives from two incentives. One was a professional interest in the phenomenon of regeneration which inevitably prompted the question of why mammals are incapable of replacing lost parts of their bodies, parts which lower vertebrates grow back so readily. While at the zoo one spring day, it dawned on me that the deer in velvet were actually regenerating histologically complex structures. When I realized how little was known about this phenomenon, it seemed appropriate to carry out experiments on growing antlers comparable to the ones previously performed on the regenerating limbs and tails of amphibians. The other reason for becoming involved was the prospect of handling large animals, a preference acquired while working on a sheep and cattle ranch in Montana one summer during my college years. What began as an avocation progressively dominated my research activities, a trend I have long since ceased to resist.

The purpose of this book is twofold. First and foremost, it is an intellectually fulfilling experience for the author. One is made abundantly aware of the gaps in one's knowledge of antler biology by such an endeavor. It has been said that ignorance, like virtue, is diminished by becoming aware of it. The more important reason is to attract larger numbers of zoologists to the field of antler research. Unsolved problems beg to be exploited, but as seems always to have been the case in the past, the number of active investigators is inadequate to the task. If this volume succeeds in arousing the interest of additional biologists in the advantages and significance of studying deer antlers, its writing will have been well worth the effort.

In the course of producing this book, I have benefited from useful discussions and correspondence with a number of colleagues more knowledgeable than I about deer and their antlers. To Anthony and George Bubenik I am especially grateful for their willingness to share their extensive knowledge with me. Zbigniew Jaczewski has been equally indispensable as a fruitful source of information over the years and, along with Richard Estes and Douglass Morse, has been good enough to review certain parts of the manuscript during the course of its gestation. Mark McNamara has generously allowed me to pick his brains on certain South American deer, for which I am grateful. Malcolm McKenna very kindly allowed me to examine specimens of fossils in the American Museum of Natural History.

Throughout the years of my research on deer, the management of the Southwick Animal Farm in Mendon, Massachusetts, has been generous in its cooperation with my needs. Among others, these include the late Justin and Daniel Southwick, and Robert, Tina, and Peter Brewer. I am grateful to Rene Pelletier and Gary Whiting for taking such excellent care of my deer in recent years. Roger Valles, director of the Roger Williams Park Zoo in Providence, has always been willing to help. Morris L. Povar, D.V.M., director of the Brown University Animal Care Facility, has been a valuable source of veterinary assistance whenever required. For their efforts in making experimental animals available for my research, I am indebted to Everett Carpenter, Charles Dana, William Davis, the late John Kennedy, Karen McAfee, Pat Quinn, Jan Smith, Phil Stanton, Norman Waycott, and many others too numerous to mention. Throughout the years, various students and assistants have helped with my research or have taken care of the deer. Among these have been Angela Black, Charlotte Clark Corkran, Charles E. Dinsmore, L. Nichols Grimes, Rebecca Cook Inman, Robert Powel, Jeffrey K. Rosen, and Susan Tilberry. Lois T. Brex has been especially dedicated to the well-being of the deer during the course of various experiments and, along with Helen Demant and Grazina Kulawas, has been of invaluable assistance in typing the manuscript. The staff of the Brown University Sciences Library have willingly risen to the challenge of obtaining even the most elusive references in the literature in a field where publications are not always in conventional sources. I am grateful to Wendy Andrews, whose graphic talents have been responsible for the excellent drawings that adorn this volume, and to Keith Vanderlin, who helped with the photography. Finally, I wish to acknowledge the National Science Foundation, the National Institutes of Health, and the Whitehall Foundation for grant support at various times during the twenty-odd year gestation of this book.

Richard J. Goss

Introduction

The wild Deer, wand'ring here & there,
Keeps the Human Soul from Care.

    William Blake, *Auguries of Innocence*

To the deer hunter, antlers are prized trophies. To the animal lover, they are magnificent ornaments adorning one of the world's most graceful animals. To the zoologist, they are fascinating curiosities that seem to defy the laws of nature. To the deer themselves, they are status symbols in the competition for male supremacy. Antlers are an extravagance of nature, rivalled only by such other biological luxuries as flowers, butterfly wings, and peacock tails. The antlers of deer are so improbable that if they had not evolved in the first place they would never have been conceived even in the wildest fantasies of the most imaginative biologists.

There are several dozen major species of deer in the world, of which all but two of the smallest possess antlers. The tiny pudu from the Andean slopes carries diminutive antlers sometimes only a few centimeters long. At the other extreme are wapiti, red deer, moose, caribou, and reindeer, whose antlers may exceed a meter in length. Antlers come in the form of simple spikes, elaborately branched racks, or palmate headpieces sometimes of impressive proportions. As secondary sex characters, antlers are

1

carried exclusively by males, except in the case of reindeer and caribou in which both sexes grow them.

Antlers are different from horns. They are branched structures, whereas horns are not. Antlers are composed of solid dead bone; the hardness of horns is created by layer upon layer of cornified epidermal cells. Antlers are renewed annually by apical growth centers. Horns are permanent structures (except in the pronghorn antelope) that grow from the base. Despite their differences, antlers and horns are analogous structures. Along with such other cephalic outgrowths as tusks and the skin-covered knobs of giraffes, they are often sexually dimorphic appendages serving as weapons and display organs in the social lives of their possessors.

The Cervidae evolved as much as 30–40 million years ago from Miocene ancestors in central Asia. Other forms, long since extinct, developed elaborately branched appendages covered with furry skin that protruded from various parts of their crania. In western North America, the merycodonts sprouted branched bony outgrowths that shed their skin, but were not themselves detached from the skull. Only deer developed the remarkable mechanism by which the dead antlers were cast off each year and replaced by new ones that are often larger and more branched than before. The selective pressures that favored the evolution of the spontaneous death of antlers, the peeling off of their dried velvet, and the casting and regeneration of the antlers themselves are unsolved problems open to considerable conjecture. Equally intriguing is the evolution of the giant Irish elk, a deer carrying antlers over 2 m long that flourished in interglacial times but died out in recent millennia. Unearthed in the hundreds from beneath the peat bogs of Ireland, this magnificent creature is a reminder of the occasional exuberance of nature that makes the study of evolution such an exciting challenge.

Whatever may have been the pathways of antler evolution, they were molded by the social functions for which these structures were designed. Their sexual dimorphism has implications for such reproductively significant behaviors as territoriality, harem formation, polygamy, and how precocious the newborn young are at birth. The presence of antlers in one or both sexes is also correlated with diet, habitat, and the herd instinct. Although antlers serve to advertise male rank, whether they do so by head-to-head combat or through the intimidation of visual display is a matter of debate. In either case, it is clear that in the course of cervid phylogeny, antlers have evolved hand in hand with the development of specific behavior patterns in various kinds of deer.

What makes antlers unique is that they are capable of replacing themselves. The growing antler, as illustrated in Figure 1, represents the most massive blastema of any regenerating system. Although antler regeneration, like that of legs and tails in amphibians, is initiated in a healing epidermal

Fig. 1. Just weeks after losing his old antlers, the wapiti sprouts new buds that are probably the world's largest regeneration blastemas. These antlers will later elongate at almost 1.75 cm/day.

wound, the generation of the original antler in the fawn or yearling proceeds in the absence of a lesion in the integument. Recent investigations have shown that the periosteum of the frontal bones is endowed with remarkable antlerogenic potentials that can be expressed even following transplantation to ectopic sites. Although the histogenesis of regenerating antlers has thus far eluded all attempts to identify the tissue(s) from which the antler bud is derived, the sequence of developmental events in the elongating antler involves chondrogenesis, osteogenesis, and *de novo* hair follicle formation. It is the hyperplastic tips of the antler tines that are responsible for the phenomenal rates of elongation that make antlers the most rapidly growing structures in the animal kingdom.

As an exception to the rule that mammalian appendages are not supposed to regenerate, antlers provide an unprecedented opportunity to observe how nature solved this problem which has frustrated experimental zoologists for so many years. There is reason to believe that the answer may lie in the mechanisms of wound healing over the amputation stump. Under exceptional circumstances the underlying cells may be deflected from the dead-end consequences of scar formation into the more productive pathways of blastema development and regeneration.

Antlers are seasonal structures dependent on the rise and fall of testosterone secretion in the male. The growth of antlers coincides with the infertility of deer in the spring and summer, when sex hormones are at minimal levels. It is the rise in testosterone secretion toward the end of the summer that causes the maturation and demise of the antler, accompanied by the loss of velvet. Experiments have shown that these hormonal fluctuations are triggered by seasonal variations in daylength. In the absence of such photoperiodic alterations, a circannual rhythm of antler replacement may be expressed. It follows that both reproductive and antler cycles in deer are profoundly affected by latitude, i.e., tropical species lack the synchronization that prevails in termperate zones.

In view of the importance of male sex hormones in antler production, castration has some interesting and exceptional effects. It causes the old antlers to be cast prematurely, permits new ones to grow (even at the "wrong" time of year), and prevents their final maturation. The result is the production of permanently viable antlers that can neither ossify nor shed their velvet. Antlers of castrated deer grow new tissue in each successive year, giving rise to tumorous growths (antleromas) that may attain monstrous proportions. These developmental caricatures, in limbo between regeneration and cancer, have not been investigated in detail.

Other peculiarities of antler growth include the occasional production of antlers by female deer, the failure of males to grow antlers at all in some cases, and a variety of abnormalities which, if antlers were grown by human beings, would have constituted a separate branch of orthopedic medicine.

Antlers are not without their medical applications. In the Orient, velvet antlers have been consumed for thousands of years for medicinal purposes. In addition to their use as an aphrodisiac, velvet antlers are prescribed along with a variety of herbs for the treatment of many medical conditions ranging from apoplexy to epilepsy. The thriving business of deer farming may generate increased scientific interest not only in the biology of antler growth but also in the controlled testing of extracts for their possible physiological activities.

Clearly, the seemingly specialized subject of deer antlers deserves more scientific attention than it has received. Unsolved problems abound, problems that may shed light on the evolution of social behavior in animals, the regeneration of lost parts in higher vertebrates, the role of hormones and nutrition in regulating growth, and the riddle of how cancer is related to normal processes of development. In the chapters that follow, the questions about antler biology may outnumber the answers. However, it is hoped that by emphasizing the gaps in our knowledge, others will be challenged to participate in a branch of the life sciences that is as enjoyable as it is important and intriguing.

General References

Brown, R. D., ed. (1983). "Antler Development in Cervidae." Caesar Kleberg Wild. Res. Inst., Texas A & I University, Kingsville.

Bubenik, A. B. (1966). "Das Geweih." Parey, Hamburg.

Chapman, D. I. (1975). Antlers—bones of contention. *Mamm. Rev.* **5,** 121–172.

Jaczewski, Z. (1981). "Poroże Jeleniowatych." Panstwowe Wydawnictwo Rolnicze i Lesne, Warsaw.

Modell, W. (1969). Horns and antlers. *Sci. Am.* **220,** 114–122.

"Records of North American Big Game" (1981). 8th ed. The Boone and Crockett Club, Alexandria, Virginia.

Ward, R. (1928). "Records of Big Game." Rowland Ward, Ltd., London.

The Diversity of Deer

Deer are even-toed ungulates, or artiodactyls. As such, they are ruminants like giraffes, pronghorn antelopes, and the Bovidae. Ruminants have the advantage of being able to eat now and chew later. Nonruminants, such as horses, possess single-chambered stomachs and must chew their food as it is eaten and swallowed. Accordingly, they must spend long hours grazing in the open. By contrast, ruminants have four-chambered stomachs. The first is the rumen which acts like a crop, enabling the animal to swallow its food without chewing as fast as it is eaten. They can therefore consume large quantitites of food in short periods of time, then repair to cover where they can chew their cud at leisure. Food is regurgitated from the rumen for mastication before it is swallowed again, this time bypassing the rumen for digestion in the other chambers of the stomach and intestines.

Ruminants have a characteristic dentition, the most notable feature of which is the lack of upper incisors. The lower ones are in occlusion with a tough fibrous upper gum against which grasses and twigs may be cropped. The lower canine teeth of deer are modified as incisors, but the upper ones are usually absent. Certain smaller species of deer, particularly those lacking antlers, possess upper canines which, in the males, may be enlarged as tusks.

The ecology and behavior of deer center around the annual reproductive cycles. During rut, the temperament of the males undergoes a dramatic reversal from the relatively benign behavior of deer in velvet to the aggressiveness of the stag in the mating season. Males at that time of year concentrate only on one thing: mating. They lose considerable weight from lack of eating, and expend their strength and energy in combat with other males and in pursuit of does. The latter come into heat every few weeks until impregnated. Depending on the species, gestation may last from 6 to 8 months, birth always being timed to occur in the spring. With few exceptions (e.g., moose calves), fawns are characteristically spotted.

The spots of a fawn are lost with its first molt in the autumn. However, adult sika deer may retain the spotted pattern in their summer coats, and fallow and axis deer may be spotted all year. Most species of deer have conspicuous rosettes of white fur around their tails. These hairs become erect when the deer is alarmed, presumably serving as a visual danger signal to others or as a device to decoy predators.

Deer are endowed with a variety of integumentary glands (Quay, 1959), the functions of which are more speculative than understood. There are facial glands in front of the eyes which in some species are developed as large, conspicuous preorbital cavities which sometimes contain waxy secretions. Between the hooves are interdigital glands, and in some species there are tarsal and metatarsal glands on the hind legs. The musk deer is famous for the scent gland on the abdomen of the male. Presumably these various glands are important for the emission of characteristic odors, especially during the mating season. Such scents, or pheromones, communicate important information between animals equipped with a correspondingly efficient sense of smell.

Although deer are not known for their visual acuity, their hearing is excellent. Large ears which can swivel in all directions complement a deer's sense of smell. This makes it exceedingly difficult to approach a deer undetected.

However, it is the sense of smell that predominates, as any experienced hunter can testify. No one has put it more vividly than Sobieski and Stuart (1848) who wrote:

> Above all things, let not the devil tempt you to trifle with a deer's nose: you may cross his sight, walk up to him in a grey coat, or, if standing against a tree or rock near your own colour, wait till he walks up to you,—but you cannot cross his nose even at an incredible distance, but he will feel the tainted air. Colours and forms may be deceptive or alike: there are grey, brown, and green rocks and stocks as well as men, and all these may be equivocal—but there is but *one* scent of *man,* and that he never doubts or mistakes; that is filled with danger and terror; and one whiff of its poison at a mile off, and whether feeding or lying, his head is instantly up,—his nose to the wind,—and in the next moment his broad antlers turn,—his single is tossed in your face, and he is away to the hill or the

wood; and if there are no green corns, peas, or potatoes in the neighbourhood, he may
not be seen on the same side of the forest for a month.

Compared with other types of animals, deer are relatively quiet creatures.
However, during the mating season, the roaring of the European red deer
and the more musical bugling of the North American elk, or wapiti, are
never-to-be forgotten sounds to anyone who has heard the autumn woods
resound with the calls of stags in rut. Other species are usually limited to
grunts, snorts, bleats, whines, and barks.

Deer are believed to have evolved originally in central Asia. From here,
their descendants migrated to Europe and the Americas. Australia and Africa
(except along part of the Mediterranean coast) have no native deer. Various
species have been widely introduced to other lands where they live in zoos,
estates, and deer parks, or have become established in the wild.

The following accounts of different types of deer are intended to highlight
the various attributes that distinguish the major species. With apologies to
taxonomists, I have taken the liberty of considering the various types of deer
more according to their geographic distribution than their taxonomic relations, although there are obvious parallels. Further, little attempt is made to
distinguish between closely related species, not because this is not important, but because for present purposes their similarities outweigh their differences. For more general sources of information on deer, the following
references are recommended: Cabrera and Yepes (1940), Cahalane (1939),
Caton (1877), Chaplin (1977), Dansie and Wince (1968–1970), de Nahlik
(1959), Flerov (1952), Lydekker (1898), Millais (1897, 1906), Phillips
(1927–1928), Prior (1965), Seton (1909), and Whitehead (1972). "Threatened Deer" (IUCN, 1978) is also recommended. The British Deer Society
publishes the journal, "Deer."

Fallow Deer
(*Dama dama*)

Fallow deer are undeniably among the world's most beautiful deer, as
their popularity in zoos and deer parks testifies (D. I. Chapman and Chapman, 1975; N. D. Chapman and Chapman, 1978). Their original range
extended throughout the Mediterranean area, and they figure prominently in
the art of near eastern countries. It is not certain whether or not fallow deer
were carried to England by the Romans, but in more recent times they have
been distributed to virtually all parts of the temperate zones both north and
south of the equator (Chapman and Chapman, 1980). The first specimens in
North America were reportedly brought to Mount Vernon by George Wash-

ington, who had hoped to cross them with the native white-tailed deer. With the possible exception of the endangered Mesopotamian deer in Iran (Haltenorth, 1959; Pepper, 1964), few if any fallow deer survive in the wild. Standing about 1 m high at the withers, the majority of fallow deer are brown with white spots, but white ones are a favorite. The latter are not albinos, as evidenced by their eye pigments and their tan color as fawns. The least common variety are the melanistic fallow which are dark brown or almost black in color. In the absence of carefully controlled breeding experiments, the inheritance of these coat colors is not well understood.

Like most other deer, fallows rut in September and October, give birth in June after a gestation period of nearly 8 months and may reach puberty at about 16 months of age. The estrous cycle of the female has been reported to be 24–26 days. Fawns show their first signs of antler production by the development of frontal pedicles in the winter and the growth of their first unbranched antlers the following spring and summer. Adults usually cast their antlers in May and lose their velvet in September.

The antlers of fallow deer are among the most spectacular headpieces carried by any of the Cervidae. Although the first few sets of antlers produced beyond the yearlings' spikes are the usual branched structures typical

Fig. 2. White fallow deer with fully grown antlers still in velvet. At this stage, massive ossification solidifies the bone and cuts off the blood supply, leading to ischemic necrosis of the skin and its eventual shedding.

of most deer, adult bucks develop palmate antlers which curve gracefully outward and upward as magnificent racks, particularly while in velvet (Fig. 2). They may grow to lengths of two-thirds of a meter and sprout a dozen points each.

Red Deer
(*Cervus elapus*)

The red deer is one of the most thoroughly investigated deer in the world (Delap, 1977), not only because of its availability to European zoologists, but because the feral descendants of red deer imported into New Zealand created an overpopulation problem in that country. This problem has now been alleviated by the popularity of deer farming (Yerex, 1979) which provides velvet antlers for medicinal purposes in the oriental market, and which meets the growing demands for venison. This economic importance has generated no small measure of interest in antler research.

The social relationships between herd members have been extensively studied in the red deer, particularly with reference to the mating behavior (Clutton-Brock et al., 1979, 1982; Darling, 1937). Rut may commence as early as late September and last well into October, although fertility may in fact persist until the end of winter. Stags roar at each other, and more evenly matched ones may engage in active combat. According to Guinness et al. (1971), females come into heat every 15–20 days (average 18 days), and single fawns are born in May or June following a gestation period of up to 8 months. Stags cast their antlers in March or April, growing new ones that reach maturity and become cleaned of velvet in late August or early September.

Red deer, sometimes carrying exaggerated antlers, have been the subjects of cave paintings by prehistoric man in Europe (Fig. 3). The red deer also serves as the prototype for trophy antlers. The popularity of hunting this magnificent animal led to the establishment of a scoring system by which record heads could be ranked. It also gave rise to a standardized nomenclature for the tines of antlers. Accordingly, the first branch is called the brow tine. The second is the bez (or bey), and the third the trez (or trey). In royal stags, the trifurcate branching of the end of the antler constitutes the crown (Fig. 4). It is this that distinguishes red deer antlers from those of the American wapiti in which the terminal branches are bifurcate.

Although red deer and wapiti are classified as separate species, they readily breed with each other. It is said that vocalization in the resultant crosses "begins with the red deer roar and ends in the wapiti whistle" (Winans, 1913). Red deer are also believed to have crossed with sika deer

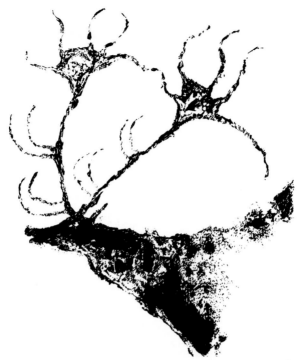

Fig. 3. Prehistoric cave painting of a red deer in France. The numbers of points and the crown have been exaggerated by the paleolithic artist. (After Leroi-Gourhan, 1965.)

since the latter's introduction into Great Britain in 1860 (Lowe and Gardiner, 1975). In certain localities, purebred individuals may no longer exist.

A variety of related species and subspecies populate much of Europe and Asia. Although the biology of these differs only in minor ways, their widespread existence testifies to the adaptability of this deer. "Isolate a small number of Red Deer in any particular open ground, park, or forest," wrote Millais (1906), "and in a few years we find a pronounced type which by hypersensitive taxonomers could be described as a new subspecies."

Sika Deer
(*Cervus nippon*)

Sika is the Japanese word for deer. Various subspecies populate Japan, Korea, Manchuria, eastern China, and Taiwan (Glover, 1956; Vidron, 1939). Its coat alternates between the dark brown pelage of winter and a reddish spotted coat in the summer. It is sometimes called the spotted deer.

Fig. 4.   Trophy specimen of a red deer. (After Millais, 1897.)

The Formosan sika is conspicuous by its bright summer coat and nearly orange-colored antlers in velvet. Dybowski's deer is a large variety of sika native to Manchuria. Because it is so easily managed and breeds so readily, the sika deer is a common inhabitant of zoos and estates. Feral populations thrive in various parts of Great Britain (Horwood and Masters, 1981). Its densest concentration is in Nara, Japan, where a large herd of relatively tame animals is a popular tourist attraction.

Rut occurs in October, and 8 months later single fawns are born in May or June. The first antlers are grown by yearlings, and are invariably unbranched spikes. Adults cast their antlers in April or May, but spike bucks may not do so until early June. The velvet is shed in early September from 4-point antlers about half a meter long (Fig. 5).

Fig. 5. The Japanese sika deer develops an 8-point head.

Although the sika deer is not as striking an animal as are some other species, it is no less endearing. When hungry, it emits a high-pitched whine, the meaning of which is unmistakable. Allowed to run free, sika deer sometimes prance on all four hooves with legs held stiff as if they were bouncing on springs.

Sambar
(*Cervus unicolor*)

This large deer stands over 1.25 m high, and inhabits the Indian peninsula from the Himalayas to Sri Lanka. Its range also extends to southeast Asia and

Fig. 6. Although twice the size of the sika, sambar antlers carry only 3 points each. (After Lydekker, 1898.)

adjacent islands. Although it has an annual molt, its coat is the same color at all seasons, as indicated by its Latin name. It has a mating season in the autumn, at least at more northern latitudes, but there is reason to believe that its reproductive cycle may in fact be irregular throughout the year, particularly among populations closer to the equator (Brander, 1944; Morris, 1934; Phillips, 1927–1928; Thom, 1937). Nevertheless, most fawns are born in May or June following an 8-month gestation period. The antler cycle may also be irregular, but where it is seasonal, antlers are cast usually in March or April and they are fully grown by September. Antlers may attain maximum lengths of 1 m and typically carry 3 points each (Fig. 6). As with other tropical deer, the sambar sometimes retains its antlers for several years in a row before replacing them.

Barasingha, or Swamp Deer
(*Cervus duvauceli*)

The barasingha is a large deer inhabiting northern India. It grows up to 1.25 m high and carries antlers almost 1 m long. These handsome racks

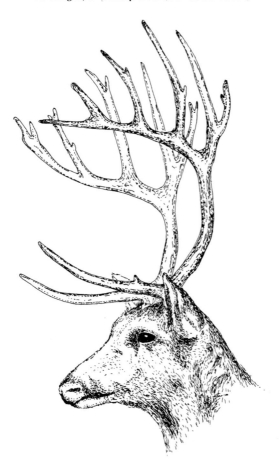

Fig. 7. The handsome barasingha deer is named for the 12 points characteristically produced by its antlers.

sweep up and forward, sprouting a series of branches that extend vertically from the distal portions (Fig. 7). Although there are reports that antler and reproductive cycles of the barasingha deer may be indefinite, especially in its native habitat, those observed in the zoos of Europe and North America tend to exhibit an autumn mating season, spring birth, and antler growth phase during the spring and early summer. The gestation period is about 8 months.

Eld's Deer, or Thamin
(*Cervus eldi*)

Eld's deer is found in parts of southeast Asia; swampy habitats are preferred. These deer are over 1 m tall and bear antlers somewhat less than 1 m in length. The most distinctive feature of the antlers is their "rocking-chair" configuration (Fig. 8). The brow tine grows almost horizontally forward in continuity with the backward arc of the main beam. Thus, the first bifurcation is like a T-joint. These animals are sometimes called "brow-antlered" deer. A number of short points adorn the apex of the beam.

The present scarcity of this deer in nature makes further observations of its biology difficult. This is unfortunate because spotty evidence suggests that

Fig. 8. Eld's deer is adorned with gracefully curved antlers.

Eld's deer has a very unusual seasonality. It is reported to give birth in October or November following a mating season in the previous March to May (Gee, 1961; Lydekker, 1898; Nouvel, 1950). Antlers are reported to have been cast in June and September, suggesting an equally unusual cycle. Why this deer (despite its habitat in the northern hemisphere) exhibits a reproductive cycle so out of phase with other deer living at similar latitudes is difficult to explain.

Schomburgk's Deer
(*Cervus schomburgki*)

Unfortunately, this deer is now extinct, and probably has been since 1932 when the last one is believed to have been shot in Thailand. Its antlers were

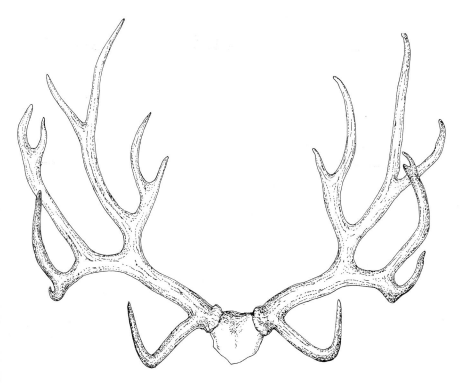

Fig. 9.   Before its extinction, Schomburgk's deer grew some of the world's most magnificent antlers.

unlike those of any other species, branching and rebranching in a dichotomous candelabra of as many as 8 points each (Fig. 9). Its range was apparently confined to restricted parts of Thailand, a fact that probably contributed to its easy extinction. When the swamps of its native habitat were converted to rice cultivation, Schomburgk's deer were forced to retreat to the forests to which they were not adapted (Giles, 1937).

First described in 1867, some specimens found their way to German zoos in the nineteenth century, but their reproduction in captivity was not sufficient to ensure the preservation of the species. Unhappily, little is known of the biology of Schomburgk's deer except for births recorded in captivity from early March to late August (Mohr, 1943). This suggests that the reproductive cycle might have been irregular.

Fig. 10. The rusa deer is the most southern representative of the genus *Cervus*. (After van Bemmel, 1949.)

Rusa Deer
(Cervus timorensis)

Throughout many of the islands of Indonesia there are various subspecies of rusa deer which, because of their nearly equatorial habitat, exhibit a wide range of breeding seasons. They stand up to 1 m high and may carry 3-point antlers almost 1 m long (Fig. 10). Their antler cycles are irregular, varying from island to island (Hoogerwerf, 1970; Valera, 1955; van Bemmel, 1949). Despite their relatively widespread distribution, there is little information available on the biology of these deer.

Hog Deer
(Axis porcinus)

This deer is well named because of its short legs and stocky build that suggest porcine proportions. It inhabits northern India and southeast Asia, where it has adapted to the tropical conditions by having lost specific breeding seasons (T. A. K., 1921). Nevertheless, populations in various localities may exhibit a degree of partial synchrony (Lydekker, 1898; Phillips,

Fig. 11. Hog deer.

1927–1928). Not surprisingly, the antler growth cycle is equally irregular. These deer are about two-thirds of a meter high with 3-point antlers one-third of a meter long (Fig. 11).

Axis Deer, or Chital
(Axis axis)

One of the world's handsomest deer, the rich brown color of its coat is punctuated with a beautiful pattern of spots that persists all year. Its natural habitat includes India and Sri Lanka, where it is one of the favorite foods of the Bengal tiger (Schaller, 1967), a fact of life that may account for the relatively high-strung temperament of this species. It is a common resident of zoos around the world, but wild populations exist in Hawaii (Graf and Nichols, 1967) as descendants of imported specimens. They also populate certain Texas ranches. Records indicate that there is little synchrony in the reproductive cycles of the axis deer, yet suggestions of latent seasonality exist (Morris, 1934; Phillips, 1927–1928). The males are capable of spirited fighting when their antlers are hard (Singhji, 1941), but the possibility re-

Fig. 12. The axis deer has conspicuous preorbital glands that give it a "4-eyed" appearance.

mains that they may be fertile throughout the year. Following a gestation period of almost 8 months, fawns can be born in any season.

Axis deer grow almost 1 m tall, with 3-point antlers three-quarters of a meter long (Fig. 12). Occasionally males may carry the same antlers for more than 1 year. Although axis deer in India reproduce and grow antlers in an irregular pattern, records of imported ones in Hawaii indicate a strong tendency to drop their antlers in December of each year. The same is true in many zoo populations. Therefore, it would seem that although the males are capable of growing antlers in any time of year, more may do so at one season than another. In any case, each individual tends to replace his antlers on a 12-month schedule, the timing of which is probably correlated with when the animal was born.

Père David's Deer, or Mi-lu
(*Elaphurus davidianus*)

The history of Père David's deer is as interesting as the animal itself. Originally native to Manchuria and eastern China, it became extinct in the wild recently enough to have been preserved in captivity. Indeed, one wonders if its near extinction might be attributed to the ancient custom of using antlers for medicinal purposes in China (Chapter 15). By the 1860s, the only surviving animals were in the Imperial Gardens in Peking where they were discovered by the French missionary—zoologist, Armand David (Delacour, 1947; Dobson, 1951). Westerners were not allowed in the Imperial Gardens, but by appropriate persuasion Father David was able to obtain some skins which were sent to Europe for examination. It was, in due course, confirmed that this was a new species, and after further persuasion (perhaps laced with a little bribery and chicanery) some live specimens were shipped to European zoos. Here they did not proliferate as well as had been hoped, and the mounting political problems in China prompted England's Duke of Bedford to buy the existing deer from continental zoos and turn them loose on his spacious estate at Woburn Abbey in Bedfordshire. In the meantime, all but one of the remaining Père David's deer in China were slaughtered when the wall of the Imperial Gardens was broken down. The remaining specimen, a female, survived without issue to the age of 21 years. In contrast, the small band of deer established and maintained in England by the Duke of Bedford and his descendants multiplied under the ideal conditions created for them there. Although there were difficulties in obtaining sufficient food during the two World Wars, the herd increased to several hundred deer by the late 1940s. It was then decided to distribute specimens to qualified zoos in other countries in order to minimize any potential risk of losing all the deer to an epidemic. Today, Père David's deer are safely ensconced in a large number of zoos throughout the world, a remarkable tribute to the dedication and foresight of the Dukes of Bedford.

This is a large deer (up to 1.25 m at the shoulder), the appearance of which is unlike any other species. It has a tail measuring 0.5 m long, including the hair, and unusually large preorbital glands. It possesses a red coat from May to September and a gray one the rest of the year. The hooves are adapted to swamplands. Its antlers are unique because the major tines branch posteriorly, instead of forward as in other species (Fig. 13). Measuring up to 0.75 m long, these antlers lack brow tines.

The mating season is in July and August when the dominant male collects practically all available females into a very impressive harem. There is reason to suspect that almost all of the progeny in a given year may in fact have the same father. Following a gestation period of 9–10 months, fawns are born in April or May. Only one other species of deer ruts in the summer and has such a long gestation period. This is the roe deer, which mates and grows antlers at the "wrong" times of year, but makes up the difference by delayed implantation. No evidence has been found that Père David's deer

Fig. 13. Père David's deer appears to have his antlers on backward.

does the same thing (Short and Hay, 1966), although they typically grow antlers in the winter as do roe bucks.

Antlers are usually cast in the late fall and new ones grow throughout the winter, reaching maturity and shedding their velvet by May or June (Bedford, 1952; Schaller and Hamer, 1978). Therefore, these deer are in velvet for approximately 6 months. In the early part of the twentieth century, when the Duke of Bedford was increasing his herd, it was noted that older males sometimes grew two sets of antlers per year (Bedford, 1952; Pocock, 1923). This phenomenon is still observed on rare occasions. One set would be grown during the autumn, the other in the winter and spring. However, under these circumstances, there seemed to be little or no indication of an extra period of rut when the first sets of antlers shed their velvet in the early winter. The explanation for this curious behavior remains to be discovered.

Muntjac
(*Muntiacus* spp.)

These secretive little deer are unique in several ways, not the least of which is the number of chromosomes in their cells. Although other deer usually have diploid numbers of 68 or 70, Reeves' muntjac has 46 and the Indian muntjac only 6 or 7 (Hsu and Benirschke, 1967–1977). How this anomalous situation arose is a matter for speculation, but it probably involved chromosomal fusion (Comings, 1971). Nevertheless, this sets muntjacs apart from other deer.

They are also distinctive in possessing very long pedicles. Because of the dark V-shaped stripes that run from the forehead along the length of the pedicles, muntjacs are sometimes referred to as rib-faced deer. The elongate pedicles are surmounted by relatively short antlers seldom more than 15 cm long (Fig. 14). They possess short brow tines, and are hooked on the ends. As if to compensate for their abbreviated antlers, male muntjacs grow upper canine tusks several centimeters long. These teeth protrude outside the lower lip, are movable in their sockets, and can be used to good advantage in fights between conspecifics or in slashing predators (Aitchison, 1946).

Various species of muntjacs grow to less than 0.5 m high, attaining weights of 15 to 25 kg. The native habitat of the Indian muntjac (*Muntiacus muntjak*) extends throughout southeast Asia from India to Vietnam and from the Himalayas to the Indonesian archipelago. The smaller Reeves' mutjac (*Mutiacus reevesi*) is native to eastern China and Taiwan.

The reproductive and antler growth cycles of the muntjac are not necessarily coordinated (Buckingham, 1981; Chaplin, 1972; Dansie, 1970; Soper, 1969). Although antlers are replaced each year during the spring and summer months as in most deer native to the temperate zone, there is no

Fig. 14.   The antlers of Reeves' muntjac seldom exceed the lengths of the pedicles on which they grow.

specific mating season. Rut occurs at any time of year, as does birth. Because females have a postpartum estrus enabling them to become pregnant right after giving birth and because the gestation period is only 6–7 months, they are potentially capable of being almost constantly pregnant thereby maximizing the frequency of fawn production. Correlated with this is the year-round fertility of the males whether they are in velvet or not. As Dansie (1970) so aptly stated, "fertility is retained despite antler casting and a female in oestrus, with a buck who has cast, does not remain unfulfilled." In their native lands, muntjacs are often called "barking deer" because of the vocalization of males, especially when a female in heat is in the neighborhood.

Tufted Deer
(Elaphodus cephalophus)

One of the least known deer in the world, this elusive animal lives in southeast China and mountainous regions of Burma and Tibet. It stands less

Fig. 15. The little-known tufted deer has the smallest antlers of any cervid.

than two-thirds of a meter high and carries antlers so short that they can only be discerned by parting the long tufts of hair on the pedicles (Fig. 15). Except for reports that its mating season may occur in the spring or summer, and that it produces one or two fawns after a 6-month gestation period, little is known about the biology of this deer (Lydekker, 1904; Yin, 1955). Like some other small Asiatic species, the males carry upper canines that are developed into articulated tusks up to 3 or 4 cm in length.

Musk Deer
(*Moschus moschiferus*)

The musk deer is unique in several ways, reflecting its taxonomic separation from other deer (Flerov, 1952). For example, they are the only deer to possess a gall bladder (an attribute shared in common with chevrotains). Whereas other deer all possess four mammary glands, musk deer have only two. Finally, males possess a single abdominal musk gland measuring up to 6 cm long and containing as much as 45 g of musk. This has been responsi-

Fig. 16. The musk deer compensates for his lack of antlers by growing an impressive pair of tusks.

ble for the severe depletion of these animals by hunters seeking musk for sale in the perfume industry.

Musk deer are native to eastern Asia where they live in mountainous regions from Siberia to the Himalayas. Their rutting period is in January, and after a 5-month gestation they give birth in June to one, sometimes two, fawns (Powell, 1964). Males lack antlers (Fig. 16), but possess upper canines developed as tusks that grow up to 10 cm long and are capable of moving forward and backwards in their sockets. These are formidable weapons which, although not used with the mouth open, are capable of inflicting wounds on rival males.

Chinese Water Deer
(*Hydropotes inermis*)

As the name implies, this little deer prefers to live near rivers in China and Korea. Standing only 0.5 m high, it is one of the two deer in the world that does not grow antlers. It is a yellow-brown color in the summer and gray in

Fig. 17. *Capreolus* means little goat, which is what the roe deer resembles.

winter. As a temperate zone animal, it has a mating season in December and after about 5.5 months gives birth usually to four fawns (sometimes as many as six) in May or June. The offspring mature rapidly, being capable of mating by the time they are 6 months old and giving birth as yearlings. The canine teeth are less than 1 cm long in females, but reach lengths of 10 cm in males owing to the capacity of these tusks to elongate continuously (Pocock, 1923).

Roe Deer
(*Capreolus capreolus*)

This relatively small deer inhabits the more northerly latitudes of Europe and Siberia where it dwells in forests and is an elusive quarry for the hunter. Its antlers point almost straight upward before giving rise to several branches at the apex (Fig. 17). The pedicles from which they arise are situated together at the top of the head, so close in fact that sometimes the two antlers may be partly fused at the midline. Trophy heads may carry antlers about 30 cm long in Europe, 45 cm long in Siberia.

The roe deer's chief claim to fame is its unorthodox type of reproduction (Bramley, 1970; Delap, 1978; Prior, 1968; "Snaffle," 1904). Most deer mate in the fall and grow antlers in the summer. The roe deer mates in the summer and grows antlers in the winter (Chapter 12). In either case, the fawns are born in the spring, usually May or early June, even though conception occurred the previous July or early August. A relatively small deer like the roe would not be expected to require a gestation period of up to 9 months, but the extra time is attributed to the delayed implantation of the early embryo (Short and Hay, 1966). This little-understood phenomenon involves an embryonic diapause starting when the blastocyst is still without a placental attachment. Thus, an event akin to suspended animation halts development but preserves the viability of the embryo while it floats within the fluids of the uterus. From late summer to early winter little or no further development occurs. However, by mid-December something triggers the restoration of growth, whereupon the embryo resumes development, implants on the uterine wall, forms a placenta, and completes its prenatal growth throughout the winter and early spring. Therefore, the effective period of gestation is closer to 5 months than 9 months.

The antler cycle is equally unusual. As with other species, the antlers are cleaned of velvet prior to the mating season, and they remain so until December. During this period, the males are extremely pugnacious and sometimes inflict serious injury on females. There is much chasing about during the courtship, often in circles 6–7 m in diameter. These roe rings, or "Hexenringe," are often centered around a tree or bush, and may sometimes describe a figure eight.

It has been claimed (Stieve, 1950) that sometimes a second rutting period may occur in the autumn, during which time females not impregnated the previous summer may be mated. This is not necessarily just a prolonged continuation of the summertime rut, but is probably a more or less separate episode. Indeed, there are cases on record (Prior, 1968; Ullrich, 1961) of roe bucks actually growing two sets of antlers per year, presumably in association with the two mating seasons.

Ordinarily, the male drops his antlers in December, or late November in older individuals. This more or less coincides with the time when belated implantation of the embryo occurs in the female. Growth of the new antlers begins as soon as the old ones have dropped off, reaching maturity in April or May when the velvet is shed. Except for Père David's deer, the roe deer is the only species that grows antlers throughout the winter months. Both of these species have also been observed to grow two sets of antlers per year (Bedford, 1952; Pocock, 1923; Prior, 1968; Ullrich, 1961). A small set of antlers is produced by fawns during the first year when they are about 6–8 months old. These first antlers grow from about November to February, when the velvet is shed, not to be replaced until the following winter when the deer is 1.5 years old.

"Snaffle" (1904) quotes Sir Harry Johnston on the subject of the anomalous reproductive cycle of the roe deer as follows:

> The explanation may possibly be that the roe originated in Eastern Asia, where its nearest relation, *Hydropotes,* still lives; and that in a semi-tropical climate there was no risk in producing young early in February, but that when pushed into the northern regions, this deer may have gradually acquired a power of retarding the development of the foetus till it could be produced at the beginning of summer.

Yet it would seem more convenient simply to have advanced the mating season from summer to fall as in other deer. Fries (1880) held that delayed implantation in the roe deer might have been an adaptation of an animal with a relatively short gestation to avoid having to mate in winter to ensure spring births. Alternatively, one could speculate that roe deer might have evolved from much larger ancestors with 9-month gestations. As their body sizes (and gestation periods) declined they could either have postponed the rutting season or invented delayed implantation. Why the latter was selected remains a mystery.

Wapiti, or American Elk
(*Cervus canadensis*)

The wapiti is the American counterpart of the European red deer, which it exceeds in size. Originally occupying the more northerly parts of North America from coast to coast, the wapiti is now largely confined to western regions where it lives in loosely organized herds that migrate to mountainous summer ranges in May and June, and back to the lowlands in November. Its summer coat is reddish, being shed in late August or early September to be replaced by a brown winter coat that molts from April to June (McCullough, 1969; Murie, 1951; Schwartz and Mitchell, 1945; Seton, 1909).

Females usually breed for the first time in their third year (approximately 2 years and 4 months of age). Although males may experience rut at the same age, they are not fully mature until they are over 3 years old. Rut occurs in September and October, and is signaled by the bugling of the males that echoes through the mountains. The males customarily wallow at this time of year, and their belligerence results in violent clashes when rivals join battle. Their antlers, which in trophy specimens may attain lengths of 1.5 m, with 12–14 points between them, make formidable weapons for such encounters (Fig. 18).

Fig. 18. The wapiti is a red deer that migrated from the Old World to North America in the Pleistocene epoch.

Birth usually occurs in late May or early June. The fawns remain spotted until the first molt in September. Although they may begin to nibble vegetation during their first month of life, fawns are not usually fully weaned until fall, or even winter. They grow their first antlers as yearlings, producing unbranched spikes that in exceptional cases may approach lengths of 0.5 m in larger animals born early in the season. Adults cast their old antlers in March or April, starting new growth immediately. The velvet is shed in August or early September, but in yearlings this may be delayed until autumn.

Moose
(*Alces* spp.)

Although there is some confusion in the name of this magnificent animal, there is no mistaking the world's largest deer. Body weights of over 600 kg are supported on long legs adapted for travel in deep snow. Larger specimens may grow to heights of 2 m. In summer, they are at home in water where they may sometimes become completely submerged in their quest for water plants. The circumpolar distribution of these animals has led to an interesting mixup in their nomenclature. Originally referred to as "elk" in Europe, this term was mistakenly applied to the American wapiti, a relative of the Eurasian red deer. Meanwhile, what the Europeans would have called an elk in the Old World was called the moose by early North American settlers. This term derived from the Algonquin Indian name for "twig-eater." Thus, in using the term elk it is important to remember which side of the Atlantic one is on.

Aside from its gigantic proportions, the moose is characterized by an oversized muzzle, the presence of a "bell" of uncertain function that may dangle 20 to 30 cm from the neck, and distinctive palmate antlers. Although the antlers grow to impressive dimensions, it is interesting to note that for an animal of this size, they are disproportionately small on an allometric basis. Record Alaskan moose grow antlers up to 1.25 m long with several dozen points and spreads of as much as 2 m (Fig. 19).

The mating season of the moose is in late September and October (Merrill, 1920; Murie, 1934; Peterson, 1955; Seton, 1909). Bull moose in rut often excavate wallows in which, for reasons best known to themselves, they instinctively cover themselves with urine and mud. So intent are they on pursuing females and fighting rivals that they may lose considerable weight from lack of eating in the early autumn. Females at this time of year come into estrus every few weeks until pregnant. Precocious ones may breed in their second year (approximately 16 months of age), but more often not until a year later. Although males may be fertile in their second year, they seldom mate until their third because of competition with older bulls. Moose reach their prime at the age of 6 to 10 years, and have been known to live up to 20 years.

Following an 8-month gestation, calves are born usually in late May or June. They weigh up to 50 kg, and are one of the few newborn deer that are not spotted. Although young mothers usually produce single offspring, older ones frequently give birth to twins, or even triplets under optimal conditions.

The young moose grows rapidly. By August its juvenile coat is shed, and

Fig. 19.   The Alaskan moose grows huge palmate antlers. Those of its European cousins tend to be smaller and more branched. (Courtesy of William E. Ruth.)

during the autumn, in more precocious calves, the first set of antlers may begin to grow. Such antlers seldom develop to more than 1–2 cm in length, and their velvet, if it is lost at all, is shed in the winter. The following spring, new antlers begin to grow, usually developing into elongate unbranched spikes in the yearlings. Two-year-old animals generally grow branched antlers, which in their third year may become palmate. Maximum dimensions may not be realized until the moose is 6 years old (Cringan, 1955).

Moose usually shed the velvet from their antlers in August or early September. After rut, the antlers themselves are dropped—in December by the oldest bulls, but at progressively later dates by younger ones. Yearlings may retain their old antlers until April or May. New antler growth usually commences in March following a 2-3-month lag from when the adult bulls cast their old antlers. Unlike most other species of deer, moose molt once per year instead of twice. Molting may begin as early as April and last through most of the summer as the previous year's coat is shed and a new one grows. By October the new coat is complete. Males and females are the same color.

Attempts to domesticate the moose have met with only limited success in Europe, Siberia, and North America. It is important to separate the calf from its mother as early as possible, and not later than 3 days after birth if it is not to acquire wild instincts (Knorre, 1974). When bottle fed, such calves become quite tame and are not afraid of man. In this way it is possible to raise cows that can be milked. Domesticated bulls are castrated, and have on occasion been trained to harness or used as pack animals.

Reindeer and Caribou
(*Rangifer tarandus*)

These animals are the most unique of all deer for a number of reasons. Their antlers are asymmetric (Chapter 10), and they are the only deer to possess antlers in both sexes. They are the most gregarious species, forming massive herds that dominate the tundra. They live farther north and migrate greater distances each spring and fall than any other deer. Finally, they have been subjected to widespread semidomestication, to the extent that reindeer have become not only the most economically useful of all deer, but are an indispensable mainstay to the livelihood of native peoples in the Arctic (Luick, 1980).

Although reindeer and caribou are normally considered members of the same species, *Rangifer tarandus*, caribou are somewhat larger than reindeer, and their temperament seems not as conducive to the type of domestication that has been achieved by the Lapps with the reindeer. The significance of this difference is emphasized by the fact that, in casting about for a suitable domestic animal in Alaska, it was worth importing reindeer from Scandinavia and Siberia instead of attempting to domesticate the native caribou (Ward, 1955, 1956). The inevitable cross-breeding has occurred.

The peak of the rutting season for reindeer and caribou occurs in October, but mating may sometimes take place in late September or early November (Dugmore, 1913; Hadwen and Palmer, 1922; Harper, 1955; Jacobi, 1931; Millais, 1907; Murie, 1935). The estrous cycle of the female has been estimated at only 10–12 days. After a gestation period of 7.5–8 months, single calves are born in May or June. They are not hidden like other fawns because the herd is in its spring migration at this time of year. The newborn calves must be ambulatory if they are to avoid being left behind to the many predators that follow the herds. Weaning occurs in September, but the young remain close to their mothers during the coming winter. Molting in reindeer and caribou is a once-a-year phenomenon. It starts in late June and continues into September as the previous year's coat is gradually shed and replaced.

Within weeks of being born, calves begin to grow their first sets of antlers. These develop into unbranched spikes during the summer and fall. They drop off in the following spring when the growth of replacements is initiated.

Fig. 20. Caribou antlers are unique in possessing brow tines that are palmate on one side and digitate on the other.

During the first few years of life the antlers become progressively larger until they reach their maximum dimensions at the age of 5–6 years in the male and about 4 years in the female. Caribou antlers (Fig. 20) have been measured at lengths of 1.5 m or more in record specimens, with up to several dozen points. Older males cast their antlers sooner than young ones, sometimes as early as late November. Less mature individuals do so at progressively later months through the winter and early spring. Therefore, the older the animal, the longer is the delay between casting and regrowth.

Females start to grow their antlers somewhat later than males, and they usually retain them longer in the winter. Barren females drop their antlers in late winter or early spring, at about the time younger males lose theirs. Pregnant ones retain them until just before or, more commonly, shortly after giving birth. Cows typically carry smaller antlers than bulls, and in some localities a percentage of the population may lack antlers altogether (Bergerud, 1976). Sometimes males also fail to produce antlers, but this is very rare. The selective advantage in the possession of antlers by females, particularly for longer periods of time by pregnant cows, is believed to give the pregnant females a greater chance of survival in competition with others in the herd (Chapter 5). This is presumably more for the benefit of her unborn young than for last year's calf, because the latter possesses antlers of its own and seems capable of getting food with or without the persistent bond with its mother.

White-Tailed, Black-Tailed, and Mule Deer (*Odocoileus* spp.)

The white-tailed deer, and its many relatives in the genus *Odocoileus*, is probably the most numerous and prolific deer in the world (Taylor, 1956). In the relative absence of natural predators in North America, its population is now held in check by hunting, disease, and starvation. Nevertheless, these most graceful of all deer abound almost everywhere in the United States as a challenging quarry for the sportsman, a lovely sight for those who appreciate wildlife, and sometimes a nuisance to the vegetable gardener. Its ancestry is obscure. Pliocene fossils have been unearthed in North America, but from what forerunners they descended remains a mystery (Chapter 4). Once the Panamanian landbridge was established in the Pliocene, white-tailed deer readily populated South America. Today, members of this genus can be found from southern Alaska to Peru, a latitudinal spread unequalled by any other species of deer.

In view of the many habitats to which *Odocoileus* has adapted, it is not surprising that numerous species, subspecies, and races have differentiated. Sir Walter Raleigh called it the Virginia deer when he first saw it there in 1584. *Odocoileus virginianus* is the most widespread species, and is generally designated the white-tailed deer in both North and South America. This species includes such subspecies as *O. virginianus borealis*, the large magnificent resident of northern forests. Farther south, the body size declines in the various subspecies inhabiting the southern states and Central America. As with many species of animals, those endemic to islands tend to be smaller than their mainland relatives. So it is that the diminutive Key deer

(*O. virginianus clavium*) typically reaches a height of less than 0.75 m and weighs up to only 50 kg (Ryden, 1978).

The mule deer (*Odocoileus hemionus*) inhabits western North America (Linsdale and Tomich, 1953). It is the largest member of the genus (Dixon, 1934a,b). Numerous subspecies have been described, the northernmost of which is the Sitka deer (*O. hemionus sitkensis*) of Alaska, a relative of the black-tailed deer (*O. hemionus columbianus*) of the Pacific northwest (Cowan, 1956). Although cross-breeding has been recorded between white-tailed deer and mule or black-tailed deer, the survival and fertility of the offspring leave much to be desired (Cowan, 1962). This may explain why these similar types of deer have preserved their identities despite overlapping habitats.

At more northerly latitudes, fawns are usually born in May or June, but births may occur later in the summer in the southern states and Mexico (Chapter 11). The gestation period approximates 200 days, but may extend 1–2 weeks on either side of this average. Although single fawns are usually produced by the youngest females, twins are more the rule for older adults, and even triplets may sometimes be produced if conditions are ideal. The fawns are born with a typical spotted coat which is retained until they are weaned at the end of the summer. It may take several years for a deer to attain its full adult size, but it is not uncommon for females to reach sexual maturity during their first year. They can become pregnant as early as 6 months of age under optimal conditions. Although males of comparable age may also be fertile, they seldom if ever have the opportunity to mate in competition with adult bucks.

Coordinated with the reproductive cycle itself are the cycles of molting and antler replacement. White-tailed deer typically have rich brown coats in the winter (sometimes with a bluish-gray tinge), alternating with a reddish pelage in the summer. The winter coat is shed in the spring, and the summer one molts in late summer to early autumn. The antlers follow a somewhat different schedule.

In mature males, the old antlers may be cast as early as December. Younger ones drop theirs later in the winter or in early spring. In either case, the loss of antlers signals the end of the fertile season. However, onset of antler growth is delayed until spring. The interval between casting and regrowth may last a couple of months, during which time the ends of the pedicles, although healed with skin, show no signs of growing antler buds. However, once growth begins elongation of the antlers proceeds rapidly, and the final size is attained toward the end of the summer. The velvet is shed in September.

The antlers of *Odocoileus* are unique. The main beam grows in a posterior direction before curving forward to produce a rack with most of the tines

Fig. 21.   The points of the white-tailed antler are unbranched outgrowths of the main beam.

pointing upward. There is considerable pearlation proximally. One of the things that makes the white-tailed deer such a challenge to the trophy hunter is that the number of points is so variable. Although 8- or 10-point bucks make splendid trophies, the largest specimens may sprout many extra tines, the presence of which adds to the scoring of the antlers for trophy purposes. Normal heads may carry antlers up to 0.75 m long with a dozen points.

White-tailed antlers are easily distinguished from those of mule and black-tailed deer. The former grow unbranched tines from the main beam (Fig. 21). The latters' antlers tend to branch dichotomously, that is, the tines of mule and black-tailed deer are themselves branched structures (Fig. 22). These deer lack true brow tines. Instead, they may possess a so-called "basal snag" that projects vertically from the medial side of the beam near its base. Whether or not this is homologous with the brow tine or the bez tine of other deer remains a matter of opinion (Pocock, 1912).

Fig. 22.   Mule deer antlers carry bifurcate branches.

The tropical relatives of the North American species have adapted to the aseasonal conditions of South America by losing the synchrony with which the members of a population go through the cycles of reproduction and antler replacement (Brock, 1965). Molting may be a continuous process throughout the year, and it is said that at lower altitudes these deer possess a summer coat at all times, although those in the Andes have permanent winter coats (Hershkovitz, 1958).

In the female, studies of *Odocoileus virginianus gymnotis* in Venezuela have shown that ovulation occurs shortly after parturition, as a result of which does may mate and become pregnant even while lactating (Brokx, 1972). Consequently, it is theoretically possible for a female to produce three fawns during the first 2.5 years of life. Therefore, in tropical latitudes white-tailed deer exhibit no specific breeding seasons, and fawns are produced at all times of the year.

The males are equally aseasonal. Although they replace their antlers every year as do temperate zone deer, this is not done in unison (Goss, 1963). Each individual grows a new set of antlers with remarkable regularity every 12 months. Consequently, about one-third of the adult male population is in velvet at any given time of year. Occasionally a male may skip one or more years, retaining the same set of antlers for a prolonged period of time. However, even in such cases, when the antlers are finally replaced it is in accordance with the original timetable. Histological examination of the testes of South American deer has confirmed that they continue to produce sperm whether their antlers are bony or in velvet. However, in the absence of controlled breeding experiments, it cannot be concluded categorically that such animals are necessarily potent and fertile while in velvet.

Recent studies of white-tailed deer in the Virgin Islands (Webb and Nellis, 1981), believed to have descended from imports to St. Croix from southeastern United States around 1790, have shown that their reproductive cycles are considerably different from those that prevail on the mainland. Although most births occur in November and December, fawns may be born in almost any month. Testicular size is maximum in late spring in most, but not all, bucks. Antlers are usually cast in October and November and new ones mature by early spring. Presumably mating occurs mostly in the spring. After nearly 2 centuries of adaptation to the climate at 18°N latitude, these deer would appear to have adopted a rather flexible reproductive cycle more typical of tropical than temperate deer.

Brockets
(*Mazama* spp.)

The brockets are small, forest-dwelling deer common to the jungles of Latin America from Mexico as far south as Paraguay and northern Argentina. Although there are several different species, among the commonest members of the genus *Mazama* are the red brockets and the brown brockets. Standing only about 0.5 m high, these deer carry unbranched antlers ranging from short spikes to outgrowths 20–30 cm long (Fig. 23). In accordance with the tropical latitudes in which they live, brockets mate, give birth, and grow antlers at all times of year. There is evidence of postpartum estrus (Gardner, 1971). Gestation is about 7 months (Vanoli, 1967).

Fig. 23. The several species of Latin American brockets grow short unbranched antlers commensurate with the relatively small size of this deer.

Pampas Deer
(*Ozotoceras bezoarticus*)

The Pampas deer lives on savannahs from Brazil to Argentina between 5° and 41° latitude. It stands about two-thirds of a meter high, and carries antlers that typically have 3 points each and reach maximum lengths of 30 cm (Fig. 24). One of the most distinctive characteristics of this deer is its strong and unpleasant smell, an odor that hardly improves the flavor of its meat. Despite this, and the fact that the deer's habits are nocturnal, it has become extremely rare and is in danger of extinction (Whitehead and de Anchorena, 1972).

Much remains to be learned about the seasonal adaptations of the Pampas deer. Like the white-tailed deer in the northern hemisphere, it is possible that Pampas deer living closer to the equator may reproduce and replace their antlers on a year-round basis (Cabrera and Yepes, 1940). Farther south there are indications of seasonality. Although the peak of the fawning season may be in the Argentine spring, and the majority of males drop their antlers in the fall and winter of the southern hemisphere, it is not uncommon for individuals to deviate considerably from this pattern.

Fig. 24. Threatened with extinction, the South American pampas deer superficially resembles the white-tail with which it may share a common ancestor.

Marsh Deer
(*Blastocerus dichotomus*)

This is a large deer slightly over 1 m high at the shoulders. It lives in southern Brazil, Paraguay, and northern Argentina where its numbers have been severely depleted. Its antlers may grow to 0.5 m in length, and typically carry 4 points in adult males as a result of bifurcations in the two main branches (Fig. 25). What scattered reports are available on the dates of antler casting and regrowth suggest that these events may not occur at any fixed season of the year (Cabrera and Yepes, 1940).

Huemul
(*Hippocamelus* spp.)

The huemul is about 1 m in height and carries 2-point antlers that are 25–30 cm long (Fig. 26). There are two species in the genus. *Hippocamelus antisensis* (the taruca) inhabits the Andes from Ecuador to northern Chile. *Hippocamelus bisulcus* lives in the southern extent of the Andes where it is honored as the national animal of Chile. No other deer lives farther south than the huemul. It migrates seasonally to higher elevations in the summer and lower ones in the winter. Like other deer in temperate zones, they mate in the fall and give birth and grow antlers in the spring.

Fig. 25. The marsh deer is the largest deer in South America.

Pudu
(*Pudu* spp.)

These rare and tiny deer, standing only about one-third of a meter high, dwell in the mountainous forests of the Andes. There are two populations, *Pudu mephistophiles* living in Ecuador and northern Peru, and *Pudu pudu* in southern Chile. As one of the world's smallest deer, the pudu carries straight unbranched antlers up to 10 cm in length (Fig. 27). Except for occasional observations on these deer in zoos, little is known of their reproductive and antler cycles in nature. Specimens imported from Chile to Germany in the 1960s, following a period of adaptation, tended to cast their antlers in December or January, and shed the velvet in May to July. Births were mostly in June following a 7-month gestation (Frädrich, 1975). Presumably this is the reverse of the seasonality in their native habitat.

Cervid Classification

The major taxonomic relationships of deer may be based on four lines of evidence. These are the fossil record, their geographic distribution, the

Fig. 26. The huemul's survival is threatened by its unwary nature, making it easy prey for hunters.

anatomy of the leg bones, and chromosome numbers. These and other characteristics related to the antler and reproductive cycles of deer may be compared from the data in Table I. Naturally, antler morphology is also important in deer classification, but the extreme variabilities to which antlers are subject makes these characters less useful than one might wish.

Fig. 27. The pudu is one of the world's smallest deer. Its antlers are correspondingly diminutive.

TABLE I

Comparison of Different Types of Deer with Respect to Their Geographic Ranges, Classification, Diploid Chromosome Numbers, and Reproductive and Antler Cycles[a]

Types of deer	Principal habitat	Plesiometa-carpalia vs. Telemeta-carpalia	Chromosome no.	Rut	Antlers cast	Birth	Velvet shed
					Months		
Fallow	South Europe	P	68	10	5	6	8–9
Red	Europe	P	68	9–10	3	5	8–9
Sika	East Asia	P	67	10	5	5–6	9
Sambar	South Asia	P	64–65	11	3–4	6	—
Barasingha	South Asia	P	56	10	2–5	6	—
Eld's	Southeast Asia	P	—	3–5	6	10–11	11
Schomburgk's	Thailand	—	—	—	—	—	—
Rusa	Indonesia	P	—	5–11	12–1	1–12	6–7
Hog	South Asia	P	68	1–12	1–12	1–12	—
Axis	India	P	66	1–12	1–12	1–12	—
Père David's	East Asia	P	68	7–8	10	4–5	5
Muntjac	South Asia	P	46, 6–7	1–12	5–6	1–12	9
Tufted	Burma, China	P	—	4–5	—	10–11	—
Musk	East Asia	T	—	1	—	6	—
Chinese Water	China, Korea	T	70	12	—	5–6	—
Roe	Eurasia	T	70	7–8	10–12	5	2–4
Wapiti	North America	P	—	9–10	3–4	5–6	8–9
Moose	Circumpolar	T	70(68)	9–10	1	5–6	8–9
Reindeer, caribou	Circumpolar	T	70	10	12	5	8–9
Mule, white-tailed	North and South America	T	70	11	1	6	9
Brocket	South America	—	68(50)	1–12	1–12	1–12	—
Pampas	South America	T	—	—	5	9	9
Marsh	South America	—	—	—	1–12	1–12	—
Huemul	South America	T	—	—	—	—	—
Pudu	South America	T	70	—	—	—	—

[a] Months (1 = January, 12 = December) represent average times of year for which specific data are available in the literature. See text for details.

Fossil remains have confirmed that deer originated in eastern Asia and later radiated to Europe, and presumably North and South America too (Chapter 4). It is no coincidence that the only kinds of deer represented on both hemispheres (moose and reindeer/caribou) inhabit high latitudes where land bridges once existed. Also, the wapiti and its close relatives in Eurasia may logically be included in this list of circumpolar deer. Otherwise, it is possible to categorize deer as either Old World or New World species.

In 1878, Sir Victor Brooke proposed that the Cervidae might be classified

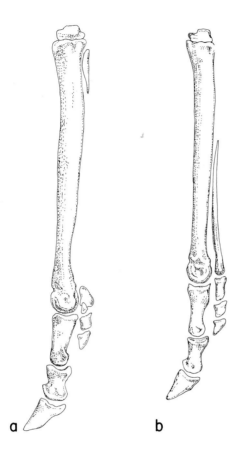

Fig. 28. Comparison of foreleg skeletal anatomy of the pleisometacarpalia (a) and tele-metacarpalia (b), showing how the second and fifth metacarpals differ in length. (After Brooke, 1878.)

according to the nature of the second and fifth metacarpal bones on their forelimbs. In all deer, the third and fourth metacarpals are fused into a cannon bone, the two hooves being correlated with the corresponding dig-its. The so-called dewclaws, relics of the second and fifth hooves, project from the leg some distance proximal to the main hooves. The associated skeletal parts extend a short distance up the leg, but in some deer there are also remnants of these bones more proximally. On the basis of these trivial but useful characters, Brooke divided the deer into two groups (Fig. 28). The Telemetacarpalia included those which possessed only the more distal ele-ments of the second and fifth metacarpals. The Plesiometacarpalia were those that retained the more proximal remnants of the second and fifth metacarpals as well. According to this classification, most of the Old World species, including the muntjac, fall in the Plesiometacarpalia. Among the exceptions are the roe deer, reindeer, and moose, which are the Tele-metacarpalia. The musk deer and the Chinese water deer, neither of which possesses antlers, are also in the Telemetacarpalia group. New World deer (except for the aforementioned wapiti) are in the Telemetacarpalia. There-fore, there exists a geographic correlation, despite a few challenging inconsistencies.

Chromosome numbers should be used as a basis for classification only with caution. However, where such information correlates with other char-acters, its credibility is enhanced. According to Hsu and Benirschke (1967–1977), the diploid number in many Old World deer in the Plesiome-tacarpalia group is 68. New World deer and the Old World Tele-metacarpalia usually have 70 chromosomes. Yet there are some disconcert-ing exceptions to this. Reeves' muntjac has 46 chromosomes, and the Indian muntjac only 6 in the female and 7 in the male (Comings, 1971). No other mammal is known to have fewer chromosomes than this. The sika deer (Cervus nippon) has 67 chromosomes (Gustavsson and Sundt, 1968), the sambar (Cervus unicolor) 64–65, and the barasingha (Cervus duvauceli) only 56. The moose may have 68 (Aula and Kaariainen, 1964) or 70 (Hsu and Benirschke, 1967–1977) chromosomes, depending on which report is accepted. The red brocket has been separately reported to have 50 (Hsu and Benirschke, 1967–1977) or 68 (Page, 1978) chromosomes. Such inconsis-tencies emphasize the dangers of relying too heavily on chromosome counts, but should not justify their total disregard, especially when they are in accord with other evidence.

Based on their lack of antlers, it would seem logical to separate the Chinese water deer and musk deer from the rest of the Cervidae. Although the former has 70 chromosomes, and both are in the Telemetacarpalia, there are no other compelling reasons to relate them to the New World deer. They would appear to have diverged from ancestral deer relatively early in cervid evolution, so much so that the musk deer is sometimes classified in a separate family (Moschidae).

Muntjacs and the tufted deer are closely affiliated. Like other Old World deer, they are in the Plesiometacarpalia, but the unusual chromosome num-bers in muntjacs, at least, argue against close affinities with other living deer. Their long pedicles and short antlers are reminiscent of early fossil deer, suggesting a link with primitive forms.

The remaining Old World Plesiometacarpalia must be assumed to be interrelated. Their anatomic and geographic consistencies override what-ever chromosomal differences may exist. Even such seemingly unrepresentative species as the fallow deer and Père David's deer cannot be unrelated because they both have 68 chromosomes and are in the Plesiometacarpalia.

Except for the wapiti, all New World deer thus far examined are in the Telemetacarpalia. Further, most have been found to possess 70 chromosomes. Therefore, on geographic, anatomic, karyologic, and developmental bases, they are conveniently distinguishable from Old World forms. The problem is to determine how the deer of the two hemispheres are related.

It is significant that both moose, reindeer, and caribou also have 70 chromosomes and are in the Telemetacarpalia. Although it is generally believed that moose originated in the Old World, it has been suggested (Frick, 1937; Flerov, 1952) that caribou may have evolved from New World ancestors and given rise secondarily to reindeer in the Old World.

In view of the fact that moose, reindeer, and wapiti successfully migrated from one hemisphere to the other, it is difficult to understand why the roe deer did not follow suit. It would have made a fine addition to the North American fauna. Its current range, although not as far north as that of the moose and reindeer, extends to higher latitudes than the red deer and wapiti (Whitehead, 1972). Like the moose and reindeer, the roe deer has 70 chromosomes (Hsu and Benirschke, 1967–1977) and is in the Telemetacarpalia. It would seem to be in the wrong hemisphere. It is tempting to speculate on the Old World origins of *Odocoileus*. Clearly, the most logical candidate as ancestor of American deer would be the roe deer. Both are in the Telemetacarpalia. Both have 70 chromosomes. However, neither is circumpolar, although together they circle the globe. It is conceivable that the ancestors of the roe deer might in fact have invaded North America and became isolated from the parent stock long enough ago to have evolved into a seemingly distinct group. Inadequate fossil evidence neither confirms nor invalidates this hypothesis. Nor are the relationships of moose and reindeer to other Telemetacarpalia understood. Speculation tends to be inversely proportional to facts, and the baffling ancestry of American deer is no exception. It can only be hoped that paleontologists of the future will rise to the challenge of unearthing the fossil evidence needed to settle this intriguing problem.

References

Aitchison, J. (1946). Hinged teeth in mammals: A study of the tusks of muntjacs (*Muntiacus*) and Chinese water deer (*Hydropotes inermis*). *Proc. Zool. Soc. London* **116**, 329–338.

Aula, P., and Kaariainen, L. (1964). The karyotype of the elk (Alces alces). *Hereditas* **51**, 274–278.

Bedford, Duke of (1952). Père David's deer. *Zoo Life* **7**, 47–49.

Bergerud, A. T. (1976). The annual antler cycle in Newfoundland caribou. *Can. Field Nat.* **90,** 449–463.

Bramley, P. S. (1970). Territoriality and reproductive behaviour of roe deer. *J. Reprod. Fertil., Suppl.* **11,** 43–70.

Brander, A. A. D. (1944). Breeding season of the Indian sambar. *J. Bombay Nat. Hist. Soc.* **44,** 587.

Brock, S. E. (1965). The deer of British Guiana. *J. Br. Guiana Mus. & Zoo.* **40,** 18–24.

Brokx, P. A. (1972). Ovarian composition and aspects of the reproductive physiology of Venezuelan white-tailed deer (*Odocoileus virginianus gymnotis*). *J. Mammal.* **53,** 760–773.

Brooke, V. (1878). On the classification of the Cervidae, with a synopsis of the existing species. *J. Zool.* 883–928.

Buckingham, W. G. (1981). Living with muntjac. *Deer* **5,** 252–253.

Cabrera, A., and Yepes, J. (1940). "Historia Natural Ediar; Mamiferos Sud-Americanos (Vida, Costumbres y Descripcion)." Compania Argentina de Editores, Buenos Aires.

Cahalane, V. H. (1939). Deer of the world. *Natl. Geogr. Mag.* **76 (4),** 463–510.

Caton, J. D. (1877). "The Antelope and Deer of America," Forest & Stream Publ. Co., New York.

Chaplin, R. E. (1972). The antler cycle of muntjac deer in Britain. *Deer* **2,** 938–941.

Chaplin, R. E. (1977). "Deer." Blandford Press Ltd., Dorset, England.

Chapman, D., and Chapman, N. (1975). "Fallow Deer." Terence Dalton Ltd., Lavenham, Suffolk, England.

Chapman, N. and Chapman, D. (1978). Fallow deer (*Dama dama*). *Br. Deer Soc. Publ.* **1,** 1–22.

Chapman, N. D., and Chapman, D. I. (1980). The distribution of fallow deer: A worldwide review. *Mamm. Rev.* **10,** 61–138.

Clutton-Brock, T. H., Albon, S. D., Gibson, R. M., and Guinness, F. E. (1979). The logical stag: Adaptive aspects of fighting in red deer (*Cervus elaphus* L.). *Anim. Behav.* **27,** 211–225.

Clutton-Brock, T. H., Guinness, F. E., and Albon, S. D. (1982). "Red Deer: Behavior and Ecology of Two Sexes." Univ. of Chicago Press, Chicago.

Comings, D. E. (1971). Heterochromatin of the Indian muntjac. *Exp. Cell Res.* **67,** 441–460.

Cowan, I. McT. (1956). Life and times of the coast black-tailed deer. *In* "The Deer of North America" (W. P. Taylor, ed.), pp. 538–545. Stackpole Co., Harrisburg, Pennsylvania.

Cowan, I. McT. (1962). Hybridization between the black-tailed deer and the white-tailed deer. *J. Mammal.* **43,** 539–541.

Cringan, A. T. (1955). Studies of moose antler development in relation to age. *In* "North American Moose" (R. L. Peterson, ed.), pp. 239–247. Univ. of Toronto Press, Toronto.

Dansie, O. (1970). "Muntjac (*Muntiacus sp.*)." Br. Deer Soc., Southampton, England.

Dansie, O., and Wince, W. (1968–1970). "Deer of the World." Br. Deer Soc., Southampton, England.

Darling, F. F. (1937). "A Herd of Red Deer. A Study in Animal Behaviour," Oxford Univ. Press, London/New York.

Delacour, J. (1947). Now we exhibit—The rarest deer in the world. *Anim. Kingdom* **50,** 2–5, 24.

Delap, P. (1977). "Red Deer (*Cervus elaphus*)," Publ. No. 5. Br. Deer Soc., Southampton, England.

Delap, P. (1978). "Roe Deer (*Capreolus capreolus*)," Publ. No. 4. Br. Deer Soc., Southampton, England.

de Nahlik, A. J. (1959). "Wild Deer." Faber & Faber, London.

Dixon, J. S. (1934a). A study of the life history and food habits of mule deer in California. Part I. Life history. *Calif. Fish Game* **20 (3),** 181–282.

Dixon, J. S. (1934b). A study of the life history and food habits of mule deer in California. Part II. Food habits. *Calif. Fish Game* **20** (4), 315–354.

Dobson, J. (1951). Père David and the discovery and early history of *Elaphurus. Proc. Zool. Soc. London* **121**, Part II, 320–325.

Dugmore, A. A. R. (1913). "The Romance of the Newfoundland Caribou. An Intimate Account of the Life of the Reindeer of North America." Lippincott, Philadelphia.

Flerov, K. K. (1952). Musk deer and deer. *In* "Fauna of USSR. Mammals." Acad. Sci. USSR, Moscow.

Frädrich, H. (1975). Notizen uber seltener gehaltene Cerviden. *Zool. Garten (Liepzig)* [N. S.] **45**, 67.

Frick, C. (1937). Horned ruminants of North America. *Bull. Am. Mus. Nat. Hist.* **69**, 1–669.

Fries, S. (1880). Uber die Fortpflanzung von *Meles taxus. Zool. Anz.* **3**, 486–492.

Gardner, A. L. (1971). Postpartum estrus in a red brocket deer, *Mazama americana,* from Peru. *J. Mammal.* **52**, 623–624.

Gee, E. P. (1961). The brow-antlered deer of Manipur. *Oryx* **6**, 103–115.

Giles, F. H. (1937). The riddle of Cervus schomburgki. *J. Siam Soc., Nat. Hist. Suppl.* **11**, 1–34.

Glover, R. (1956). Notes on the sika deer. *J. Mammal.* **37**, 99–105.

Goss, R. J. (1963). The deciduous nature of deer antlers. *In* "Mechanisms of Hard Tissue Destruction" (R. Sognnaes, ed.), Publ. No. 75, pp. 339–369. Am. Assoc. Adv. Sci., Washington, D.C.

Graf, W., and Nichols, L., Jr. (1967). The axis deer in Hawaii. *J. Bombay Nat. Hist. Soc.* **63**, 629–734.

Guinness, F., Lincoln, G. A., and Short, R. V. (1971). The reproductive cycle of the female red deer, *Cervus elaphus* L. *J. Reprod. Fertil.* **27**, 427–438.

Gustavsson, I., and Sundt, C. O. (1968). Karyotypes in five species of deer (*Alces alces* L., *Capreolus capreolus* L., *Cervus elaphus* L., *Cervus nippon nippon* Temm. and *Dama dama* L.). *Hereditas* **60**, 233–248.

Hadwen, S., and Palmer L. J. (1922). Reindeer in Alaska. *U. S., Dep. Agric., Bull.* **1089.**

Haltenorth, T. (1959). Beitrag zur Kenntnis des Mesopotamischen Damhirsches-*Cervus* (*Dama*) *mesopotamicus* Brooke, 1875—und zur Stammes und Verbreitungsgeschichte der Damhirsche allgemein. *Saugetierkd. Mitt.* **7**, 1–89.

Harper, F. (1955). "The Barren Ground Caribou of Keewatin," University of Kansas, Mus. Nat. Hist., Misc. Publ. No. 6. Allen Press, Lawrence.

Hershkovitz, P. (1958). The metatarsal glands in white-tailed deer and related forms of the neotropical region. *Mammalia* **22**, 537–546.

Hoogerwerf, A. (1970). "Udjung Kulon: The Land of the Last Javan Rhinoceros." Brill, Leiden.

Horwood, M. T., and Masters, E. H. (1981). "Sika Deer (*Cervus nippon*) (with particular reference to the Poole Basin)," Publ. No. 3 Br. Deer Soc., Southampton, England.

Hsu, T. C., and Benirschke, K. (1967–1977). "An Atlas of Mammalian Chromosomes," Vol. 10, Folios 451–517. Springer-Verlag, Berlin/New York.

International Union for Conservation of Nature and Natural Resources. (1978). "Threatened Deer," Proc. Working Mtg. Deer Specialist Grp. Survival Serv. Comm., 1977. IUCN, Morges, Switzerland.

Jacobi, A. (1931). Das Rentier. *Zool. Anz.* **96**, 1–264.

Knorre, E. P. (1974). Changes in the behavior of moose with age and during the process of domestication. *Nat. Can.* **101**, 371–377.

Leroi-Gourhan, A. (1965). "Préhistoire de l'art occidental." Editions d'Art Lucien Mazenod, Paris.

Linsdale, J. M., and Tomich, P. Q. (1953). "A Herd of Mule Deer." Univ. of California Press, Berkeley.

Lowe, V. P. W., and Gardiner, A. S. (1975). Hybridization between red deer (*Cervus elaphus*) and sika deer (*Cervus nippon*) with particular reference to stocks in N. W. England. *J. Zool.* **177,** 553–566.

Luick, J. R. (1980). Circumpolar problems in managing populations of wild and domestic reindeer. "Proceedings of the Second International Reindeer/Caribou Symposium" (E. Reimers, E. Gaare, and S. Skjenneberg, eds.), pp. 686–688. Direktoratet for vilt og ferskvannsfisk, Trondheim.

Lydekker, R. (1898). "The Deer of All Lands. A History of the Family Cervidae Living and Extinct." Rowland Ward, Ltd., London.

Lydekker, R. (1904). The Ichang tufted deer. *Proc. Zool. Soc. London* **2,** 166–169.

McCullough, D. R. (1969). The Tule Elk. *Univ. Calif., Berkeley. Publ. Zool.* **88,** 1–209.

Merrill, S. (1920). "The Moose Book." Dutton, New York.

Millais, J. G. (1897). "British Deer and Their Horns." Henry Sotheran & Co., London.

Millais, J. G. (1906). "The Mammals of Great Britain and Ireland," Vol. 3, pp. 71–179. Longmans, Green, London.

Millais, J. G. (1907). "Newfoundland and Its Untrodden Ways." Longmans, Green, London.

Mohr, E. (1943). Die ehemalige Hamburger Zucht des Schomburgk-Hirsches, *Rucervus schomburgki* Blyth. *Zool. Anz.* **142,** 30–35.

Morris, R. C. (1934). Growth and shedding of antlers in sambar (*Rusa unicolor*) and cheetal (*Axis axis*) in south India. *J. Bombay Nat. Hist. Soc.* **37,** 484.

Murie, A. (1934). The moose of Isle Royale. *Univ. Mich. Mus. Zool., Misc. Publ.* **25,** 1–44.

Murie, O. J. (1935). "Alaska-Yukon Caribou," North Am. Fauna No. 54. U. S. Dept. Agric., Washington, D.C.

Murie, O. J. (1951). "The Elk of North America." Stackpole Publ. Co., Harrisburg, Pennsylvania.

Nouvel, J. (1950). Note sur la reproduction du Cerf d'Eld (*Rucervus eldi* Guthrie) au Parc Zoologique du Bois de Vincennes. *Bull. Mus. Hist. Nat. (Paris)* **22,** 682–683.

Page, F. J. T. (1978). Review of "Atlas of Mammalian Chromosomes," by T. C. Hsu and K. Benirschke. *Deer* **4,** 332.

Pepper, H. J. (1964). The Persian fallow deer. *Oryx* **7,** 291–294.

Peterson, R. L. (1955). "North American Moose." Univ. of Toronto Press, Toronto.

Phillips, W. W. A. (1927–1928). Guide to the mammals of Ceylon. Part VI. Ungulata. *Ceylon J. Sci., Sect. B* **14,** 1–50.

Pocock, R. I. (1912). On antler growth in the Cervidae, with special reference to *Elaphurus* and *Odocoileus (Dorcelaphus)*. *Proc. Zool. Soc. London* pp. 773–783.

Pocock, R. I. (1923). On the external characters of *Elaphurus, Hydropotes, Pudu,* and other Cervidae. *Proc. Zool. Soc. London* pp. 181–207.

Powell, A. N. W. (1964). The musk deer. *J. Bengal Nat. Hist. Soc.* **33,** 149–152.

Prior, R. (1965). "Living With Deer." Andre Deutsch Limited, London.

Prior, R. (1968). "The Roe Deer of Cranborne Chase." Oxford Univ. Press, London/New York.

Quay, W. B. (1959). Microscopic structure and variation in the cutaneous glands of the deer, *Odocoileus virginianus. J. Mammal.* **40,** 114–128.

Ryden, H. (1978). Saga of the toy deer. *Audubon* **80** (6), 92–103.

Schaller, G. B. (1967). "The Deer and the Tiger. A Study of Wildlife in India." Univ. of Chicago Press, Chicago.

Schaller, G. B., and Hamer, A. (1978). Rutting behavior of Père David's deer *Elaphurus davidianus. Zool. Garten (Leipzig)* [N. S.] **48,** 1–15.

Schwartz, J. E., and Mitchell, G. E. (1945). The Roosevelt elk on the Olympic Peninsula, Washington. *J. Wildl. Manage.* **9,** 295–319.

Seton, E. T. (1909). "Life-Histories of Northern Animals. An Account of the Mammals of Manitoba," Vol. 1. Scribner's, New York.

Short, R. V., and Hay, M. F. (1966). Delayed implantation in the roe deer *Capreolus capreolus*. *In* "Comparative Biology of Reproduction in Mammals" (I. W. Rowlands, ed.), pp. 173–194. Academic Press, New York.

Singhji, M. (1941). The spotted deer. *J. Sind Nat. Hist. Soc.* **5,** 5–9.

"Snaffle," D. R. (1904). "The Roedeer." E. M. Harwar, London.

Sobieski, J., and Stuart, C. E. (1848). "Lays of the Deer Forest." William Blackwood, Edinburgh.

Soper, E. (1969). "*Muntjac*." Longmans, Green, London.

Stieve, H. (1950). Anatomische Untersuchungen über die Fortpflanzungstätigkeit des europäischen Rehes (*Capreolus capreolus capreolus* L.). *Z. Mikrosk. Anat. Forsch.* **55,** 427–530.

T. A. K. (1921). A baby hog deer in captivity. *J. Bombay Nat. Hist. Soc.* **28,** 271–273.

Taylor, W. P., ed. (1956). "The Deer of North America." Stackpole Co., Harrisburg, Pennsylvania.

Thom, W. S. (1937). The Malayan or Burmese sambar (*Rusa unicolor equinus*). *J. Bombay Nat. Hist. Soc.* **39,** 309–319.

Ullrich, W. (1961). Zweimalige Geweihbildung in Jahresablauf bei einem Rehbock. *Zool. Garten (Leipzig)* [N. S.] **25,** 411–412.

Valera, R. B. (1955). Observations on the breeding habits, the shedding and the development of antlers of the Philippine deer (*Rusa sp.*). *Philipp. J. For.* **11,** 249–257.

van Bemmel, A. C. V. (1949). Revision of the Rusine deer in the Indo-Australian archipelago. *Treubia* **20,** 191–262.

Vanoli, T. (1967). Beobachtungen an Pudus, *Mazama pudu* (Molina, 1782). *Saugetierkd. Mitt.* **15,** 155–163.

Vidron, F. (1939). "Le Cerf Sika." Paul Lechevalier, Paris.

Ward, K. (1955). A study of the introduction of reindeer in Alaska. *J. Presbyt. Hist.* **33** (4), 229–237.

Ward, K. (1956). A study of the introduction of reindeer in Alaska—II. *J. Presbyt. Hist.* **34** (4), 245–256.

Webb, J. W., and Nellis, D. W. (1981). Reproductive cycle of white-tailed deer of St. Croix, Virgin Islands. *J. Wildl. Manage.* **45,** 253–258.

Whitehead, G. K. (1972). "Deer of the World." Viking Press, New York.

Whitehead, G. K., and de Anchorena, M. (1972). Operation pampas deer. *Country Life* **152,** 596.

Winans, W. (1913). "Deer Breeding for Fine Heads with Description of Many Varieties and Cross-breeds." Rowland Ward, London.

Yerex, D. (1979). "Deer Farming in New Zealand." Deer Farm. Serv. Div. Agric. Prom. Assoc., Wellington.

Yin, U. T. (1955). Tufted deer in Burma (*Elaphodus cephalophus* Milne Edwards). *J. Bombay Nat. Hist. Soc.* **53,** 123–125.

Horns and Tusks

Status symbols are as important to animals as they are to people. Not uncommonly, they take the form of cephalic outgrowths which have evolved in a wide variety of creatures (Davis, 1974; Geist, 1966a) (Fig. 29). For example, the rhinoceros beetle is endowed with impressive protuberances from its head and thorax which are presumably analogous to the horns of higher forms. Among reptiles, the male Jackson's chameleon possesses a trio of horns enveloped in a cornified sheath reminiscent of those carried by the extinct *Triceratops*. In the rhinoceros the horn is composed of longitudinally oriented hairlike tubules of keratin matted together into a hard and formidable nosepiece. Giraffes and okapis possess bony protuberances on their skulls that remain covered with skin and hair. More conventional horns consist of a bony core enveloped in a sheath of heavily cornified epidermis. They differ from antlers in that they are unbranched, nondeciduous structures whose hardness derives from the permanent integument rather than the underlying bone.

Not surprisingly, some animals have modified their teeth into hornlike structures. The tusks of elephants are remarkable examples of dental growth potentials. In the babirusa, a wild pig of Sulawesi, males possess curved canine tusks that just miss penetrating the skull in the trajectory of their

Fig. 29. Note the diversity of horns, and the creatures that grow them, such as (a) the rhinoceros beetle, (b) Jackson's chameleon, and (c) *Triceratops*.

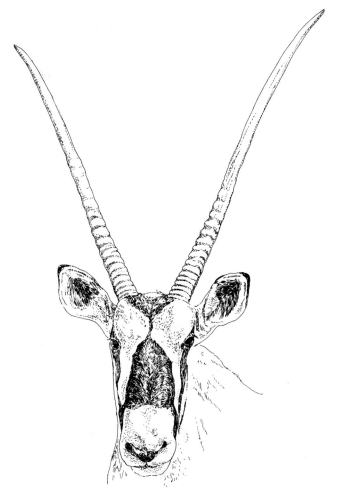

Fig. 30.   The gemsbok (*Oryx gazella*) of southern Africa has remarkably straight horns.

growth (MacKinnon, 1981). The magnificent tusks of the male narwhal are also instruments of aggression during the mating season.

Horns are of importance in the consideration of antlers for two reasons. One is that the uses to which they are put are often analogous to the functions of antlers. The other is that they emphasize, in their modes of development and final anatomy, evolution's many and varied experiments in the production of cephalic appendages.

Horns exist in many forms. The simplest configuration is exemplified by

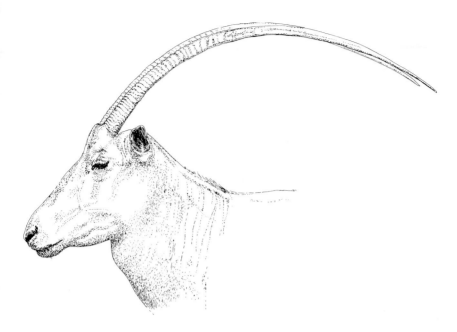

Fig. 31. The scimitar-horned oryx (*Oryx dammah*), from the fringes of the Sahara desert, carries magnificently curved horns.

such animals as the gemsbok, whose horns are long, straight, rapierlike weapons (Fig. 30). In the sable and scimitar horned antelope, they bend gracefully backwards from the head (Fig. 31). Those of the Dama gazelle do likewise before curving upward at the ends (Fig. 32). In bighorn sheep, the horns also curve backward, but continued growth may bring them to full circle on either side of the head. Domestic breeds of rams may complete two revolutions (Fig. 33). The horns of cattle tend to curl laterally. In the springbok, they grow vertically from the head before curving inward. The gnu carries horns that grow laterally downward and then hook up again at the ends. The most graceful horns are those of the kudu and Indian black buck (Fig. 34) whose growth traces a spiral trajectory. In the Himalayan markhor (Fig. 35), the horns are flattened in cross section, but twist in a helical configuration. There seem to be few potential geometries that have not been adopted by one kind of ungulate or another.

The developmental mechanics involved in horn growth presumably results from differential rates of proliferation at the base. Faster growth on one side would tend to deflect the horn in the opposite direction in much the same way one steers a rowboat. The morphogenetic pattern of a horn is

Fig. 32. In the Dama gazelle (*Gazella dama*), the even spacings between ridges reflect the precision with which horn growth is regulated.

Fig. 33. The Dorset ram grows horns that circle dangerously close to the snout, sometimes describing two complete revolutions.

therefore programmed into the germinative tissues at the base to yield a spatial and temporal pattern of faster or slower proliferation such that the prescribed horn structure is extruded.

How Horns Grow

Antlers grow at their tips, like the branches of a tree. Horns elongate like grass, growing from the base. The bony horn core is enveloped in a layer of skin, the epidermis of which differentiates into an outer cornified sheath. The earliest horn bud of a young animal is pushed distally as new material is added at the base. Both the bony core and the cornified sheath elongate during the course of maturation. The basal layer of epidermis, which retains proliferative potential, differentiates layer upon layer of keratinized tissue throughout the length and circumference of the conical bony core. In this manner, horny lamellae are added internally, in the nature of a cone within a cone (Fig. 36). This process results in the distal displacement of material produced earlier. In microscopic sections of the horn sheath, thin lamellations are visible in the cornified epidermis. It is possible that these may

Fig. 34.   The gracefully turned horns of the Indian black buck (*Antilope cervicapra*) can be traced in successive stages in the same animal over a span of 30 months.

represent the daily increments of epidermal growth, analogous to the similar lamellae visible in the enamel and dentine of teeth.

Because of the rigid nature of the horn sheath, it is impossible to modify its shape once it has been produced. Therefore, a horn can only change form by altering the mechanics of growth at the base. The tapered configuration of horns results from augmentation of the previously formed sheath with successive generations of internally deposited cones of keratinized epidermis. This has the effect of widening the basal circumference as elongation continues.

The rates at which horns elongate change with age, are affected by the seasons, and vary from animal to animal. Although horn buds are palpable

Fig. 35. The markhor (*Capra falconeri*) carries spectacular horns that coil upward in an ostentatious display of cornified elegance.

in most animals at birth, elongation may not begin for several weeks or months in an animal such as the bighorn sheep (Hansen, 1965). Males of this species grow horns almost 1 m long, adding as much as 22 cm/year during the early years of life when the rate of elongation is at its peak. This is equivalent to increments of 1 mm/day (Hemming, 1969). Although growth

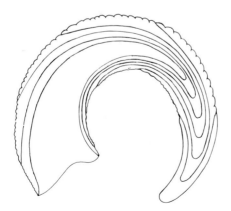

Fig. 36. Diagrammatic representation of how horns grow in the bighorn sheep. Each annual increment is laid down inside previous ones, pushing the latter ahead of them.

is sustained throughout life, it diminishes to only a few centimeters per year after maturity at about 5 years of age (Murie, 1944).

In the bighorn sheep, age determination is facilitated by the annual rings that separate one year's horn increment from the next (Geist, 1966b; Welles and Welles, 1961). These rings are an accurate reflection of the animal's age because they record how much horn was added each year throughout the animal's life. A similar pattern prevails in most horned mammals native to the temperate zone. During their reproductive season the elongation of their horns is held in abeyance. In the bighorn sheep, this period extends from about November to February or March. The rutting period is in November and December, when rams engage in their famous ritual of battering their horns together. It would seem only logical not to risk injury to the germinative tissues of the horn while the testosterone levels are elevated. Therefore, horn growth (like antler development) is reserved for the infertile phase of the year, a period of about 7 months. A similar situation is encountered in other horned ungulates in the temperate zone, but none exhibits such conspicuous annual demarcations in horns as does the bighorn sheep. It is this yearly cessation of growth that leaves its mark where elongation left off one year and began the next.

Tropical animals do not exhibit annual rings in their horns. This may be because of the year-long growth of the horns in such animals. Age is not easily determined in tropical bovids, not only because their horns grow continuously, but also because horn elongation tends to cease altogether after maturity (Roettcher and Hofmann, 1970; Spinage, 1967).

Whether annual increments of horn growth are present or not, many bovids possess horns with lesser rings that adorn their surfaces. These corrugations presumably provide traction when males lock horns. They may vary from large bumps on the surface of the horn, as in those of the ibex (Fig. 37) (only a few of which are produced each year), to smaller and more frequently formed ridges, as in the bighorn sheep (Fig. 38). An average of about 13 of these ridges are typically formed between successive annual rings in the bighorn, suggesting that a little over 2 weeks may be required for the development of each one. It may or may not be a coincidence that this approximates the estrous cycle in the ewe. The rates at which comparable rings are laid down in other species, including those from tropical habitats, varies considerably. In most forms, horn growth is maximal during the period of maturation, slowing down or stopping altogether later on (Roettcher and Hofmann, 1970; Spinage, 1967). Such rings could be used for age determination if enough were known about the growth rates of horns in each species and the age at which elongation normally ceased.

Fig. 37. The horns of the Nubian ibex exhibit conspicuous rugosities, even the largest of which represent only localized thickenings of keratin in the horn sheath. Although their mode of development has not been investigated, their initiation is associated with a prominent swelling on the animal's brow (arrow).

Fig. 38. In the Dall sheep (*Ovis dalli*) the sweeping curvature of the horns is punctuated with numerous corrugations between the less frequent yearly growth rings.

The Improbable Pronghorn

Francisco Vasquez de Coronado was probably the first European to see the pronghorn antelopes that abounded on the Great Plains west of the Mississippi (Leister, 1932). In 1804, Lewis and Clark were sufficiently impressed by this animal that they brought back a skin for museum display, no doubt after having consumed the rest of the carcass. However, not until the middle of the nineteenth century did accounts begin to trickle in that the pronghorn might shed its horns each year (Lyon, 1908). No less an authority than John James Audubon himself was inclined not to believe the tales hunters told that America's only antelope replaced its horns annually. Even after this phenomenon had been documented by Bartlett in 1865, based on his observations of a specimen in the London Zoological Gardens, skepticism lingered that such an incredible event could be true. After all, no other ungulate in the world shed its horns; and only deer antlers were supposed to drop off and regenerate.

The pronghorn antelope (*Antilocapra americana*) (Fig. 39) is the sole survivor of numerous ancestors that once roamed North America since the Miocene epoch. Some of them must have carried strange-looking horns,

Fig. 39.   The branched configuration of the pronghorn is an attribute of the horn sheath, not the underlying bony core.

judging from fossil remains, the sheaths of which may have been deciduous like those of their only remaining descendant. The method by which the pronghorn sheds its horn sheaths is as interesting as it is pertinent to the problem of antler regeneration (Caton, 1876; O'Gara and Matson, 1975).

As is the case with antlers, the horn sheaths of the pronghorn are shed after the mating season (O'Gara et al., 1971). Rut occurs in September and the horn sheaths are lost anywhere from October to December in males, usually later in females (O'Gara, 1969). Kids grow horns a few centimeters long during their first summer and fall, shedding them in January (Lyon, 1908). As yearlings they will produce prongless horns up to 15 cm long.

If one examines the inside of the shed sheaths, numerous hairs may be seen protruding from the inner surface, hairs that are embedded in the cornified matrix. These would appear to be the last structures by which the old sheath was anchored on (Bailey, 1920; Noback, 1932; Skinner, 1922).

Following shedding, the new horn is revealed at the tip of the bony core. This hairless, rubbery structure soon hardens (Skinner, 1922). More proximally, the shaft of the horn is covered with numerous hairs.

New growth begins in the winter and continues until summer by the usual buildup of keratinized layers that may rise to heights of 25 cm in males, considerably less in females. Hairs typically become embedded in the developing horn sheath (Grinnell, 1921). Although the horn of this animal appears branched, the prong is produced solely by the buildup of keratin on the anterior side of the tapered, unbranched bone core. During subsequent elongation of the horn, the prong is displaced upward, coming to lie distal to the bone core in the fully developed horn (Fig. 40).

The chief difference between the pronghorn and other horned animals is that the former sheds its sheath annually. It is interesting to note that if the pronghorn antelope is castrated, he is prevented from detaching one sheath from another (Pocock, 1905). Accordingly, annual increments of horny material accumulate giving rise to an abnormally curved structure (Fig. 41) superficially resembling what naturally occurs in other horned animals. However, castration is not without its effects on the horns of other ungulates. For example, in the impala, it is said to cause the horns to be abnormally shaped and to prevent the development of rings (J. D. Skinner, personal communication, 1972).

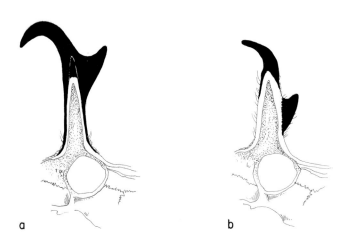

a   b

Fig. 40. In November (a), the old sheath (black) of the pronghorn obscures the incipient new one on the apex of the bone core. After the former is shed, renewed deposition of cornified tissue gives rise to the curved apical segment separately from the anterior prong, as seen in January (b). The prong is pushed distally as development progresses throughout the spring and summer. (After O'Gara and Matson, 1975.)

Shedding of the horn sheath by pronghorn antelopes is reminiscent of other molting phenomena, such as the loss of nuptial plumage by birds after their mating seasons. In the puffin, the outer sheath of its colorful beak is shed and replaced each year. In the Dalmatian pelican, the male carries a conspicuous keratinized keel on its upper beak, a structure that is cast and replaced annually. These various examples of massive loss and regeneration of tissue is nature's way of replacing damaged secondary sex characters each year in time for the next mating season.

Even in ungulates other than the pronghorn antelope, the loss of keratinized tissue from the horns is not impossible (O'Gara and Matson, 1975). The tips of the horns in bighorn sheep often become frayed from abrasion, a phenomenon known as "brooming." Exfoliation of horny material in bovids is not uncommon, and the outer sheath of the horn of the nilgai, India's largest antelope, sometimes cracks and peels off. In the wisent, or European bison, the outermost sheath of horn may be lost occassionally, revealing a shiny new horn beneath. These phenomena may represent extensions of the normal process by which prenatally developed keratinized tissue is shed after birth, or of the replacement of the horn sheath at puberty in certain horned animals.

Fig. 41. Castrated pronghorn with successive horn sheaths unshed. (After Pocock, 1905.)

The Histogenesis of Horns

In order to identify the tissues of origin from which horns arise, deletion and transplantation experiments have been most revealing. The simplest operation is the most effective, namely, to remove the skin from the presumptive horn site on the heads of young animals. This operation typically prevents horn formation, and is an effective method of dehorning animals (Brandt, 1928; Dove, 1935; Kômura, 1926). The application of caustic potash, or subcutaneous injections of concentrated solutions of calcium chloride (Koger, 1976), both of which cause extensive necrosis of the skin, effectively interfere with normal horn development. In contrast, if the underlying connective tissue, periosteum, or bony protuberance from the skull is removed, subsequent horn development is not prevented (Dove, 1935; Kômura, 1926; Marchi, 1907).

If the skin normally destined to give rise to a horn is transplanted elsewhere on the head, a correspondingly ectopic horn is produced at the graft site, while none develops in the normal location (Brandt, 1928; Dove, 1935; Kômura, 1926; Marchi, 1907). Again, this provides compelling evidence that the horn is derived exclusively from the skin. Indeed, such opera-

Fig. 42. Ayershire bull in which the two horn buds were surgically fused in the calf, resulting in the production of a single horn as the animal matured. (After Dove, 1936.)

tions enabled Dove to create his famous unicorn at the University of Maine in the 1930s. He grafted the skin from two horn sites in a calf to the midline of the head, whereupon the two potential horns developed as a single fused structure on the midline as shown in Figure 42 (Dove, 1936).

Although the horn sheath is a derivative of the skin, it is associated with a bony core firmly adherent to the skull. In young bovids, the *os cornu* originates in the subcutaneous connective tissue and fuses secondarily with the underlying skull. Vertically oriented trabeculae may develop in the cranium in association with horn development (Dürst, 1902). Cartilage is not known to be present in the developmental stages of the horn bone (Fambach, 1909). Because this bone develops as an independent center of ossification (Fig. 43), it is an epiphysis rather than an apophysis as in deer antlers. The fact that such bone development fails to occur when the overlying presumptive horn skin has been removed, yet does develop in association with grafts of such skins elsewhere on the skull, suggests that it may be induced by, or derived from, the integument.

The fur-covered horns of the giraffe develop in a similar manner (Fig. 44). Here the *os cornu*, or ossicone, can be seen in fetal animals at the site where the horn will later develop (Lankester, 1907; Spinage, 1968). Histological examination of these tissues strongly suggests that the ossicone arises in the connective tissue of the skin rather than in the periosteum of the skull. In its early stages of development it is histologically separate from the latter. The ossicone originates as a cartilaginous structure in the fetus and secondarily fuses with the parietal and, later, frontal bones of the cranium, a process that may not be complete until after the male giraffe's fourth year. In the meantime, ossification occurs at its upper end, eventually leaving only a basal

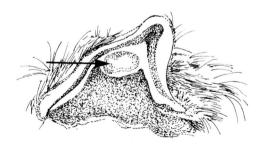

Fig. 43. *Os cornu* (arrow) as seen in sagittal section through the horn bud of a young goat. It lies between an elevation on the frontal bone and the overlying integument. (After Fambach, 1909.)

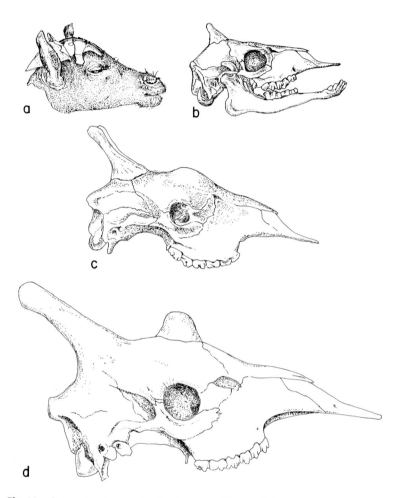

Fig. 44. Successive stages in the development of the giraffe horn, as seen in (a) the fetus, (b) postnatal infant, (c) a two-thirds grown young, and (d) adult. The position of the ossicone is indicated by an oval on the parietal bone in the fetus from which the scalp and periosteum have been reflected. The ossicone does not ankylose to the skull until adulthood. (After Lankester, 1907.)

layer of cartilage which disappears after fusion to the skull. Further growth of the horn occurs by appositional ossification. The overlying skin retains its fur, except on the upper end where it develops a thickened pad of epidermis. Because of the mechanism by which the giraffe horn develops, its bony component is logically classified as an epiphysis. In view of the unlikely possibility that deletion or transplantation experiments will ever be carried out on young giraffes, we may never learn if the primary impetus for horn development in this animal resides in the ossicone or its overlying skin.

Tusks

The mechanisms by which tusks and horns grow have much in common, despite the fact that they are entirely different structures. Tusks are teeth, and as such they consist of enamel and dentine laid down respectively by ameloblasts, specialized epidermal cells, and odontoblasts, mesodermal cells in the pulp cavity. Tusks are continuously erupting teeth. They are therefore analogous with the incisors of rabbits and rodents, teeth which grow incessantly throughout the life of the animal. In the rodent incisor, occlusion with the opposite tooth tends to reduce the rate of eruption by about one-half of the maximal potential. When the opposite tooth is removed, eruption accelerates accordingly. If, as a result of malocclusion, these incisors cannot be abraded, their continuous growth results in abnormally elongated tusks which may in due course interfere with feeding and bring about starvation.

The tusks of elephants, various wild pigs, and narwhal, and certain of the smallest species of deer are normally unoccluded. It is worth noting that the elephant tusk and the mouse incisor both erupt at a rate of about 0.4 mm/

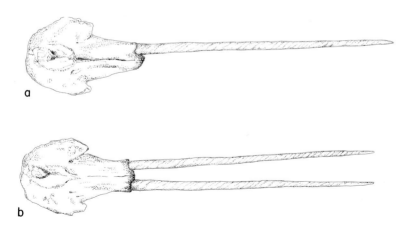

Fig. 45. Narwhal skulls showing the normally left-sided development of the male tusk (a) and the rare and unusual bilateral production of tusks (b). In either case, the tusks spiral to the animal's left.

day (Ness, 1965). Perhaps the most interesting of all tusks is that carried by the male narwhal (*Monodon monoceros*). This tusk, used in combat by males during the mating season, may grow 2 m long, or over one-half the length of the animal's body. It is derived from the left upper canine tooth of the male, and is engraved with spiral ridges that always twist in a left-hand turn. In extremely rare circumstances, both of the narwhal's upper canine teeth may develop into tusks. Although this may restore a degree of symmetry to the animal, it is a curious fact that both of these tusks still spiral to the left (Fig. 45). The mystery of its morphology is as interesting as its legendary uses as a unicorn horn (Chapter 15).

References

Bailey, V. (1920). Old and new horns of the prong-horned antelope. *J. Mammal.* **1**, 128–129.

Bartlett, A. D. (1865). Remarks upon the affinities of the prongbuck (*Antilocapra americana*). *Proc. Zool. Soc. London* pp. 718–725.

Brandt, K. (1928). Die Entwicklung des Hornes beim Rinde bis zum Beginn der Pneumatisation des Hornzapfens. 17. Beitrag zum Bau and zur Entwicklung von Hautorganen bei Säugetieren. *Morphol. Jahrb.* **60**, 428–468.

Caton, J. D. (1876). The American antelope or prong buck. *Am. Nat.* **10**, 193–205.

Davis, T. A. (1974). Of horns, antlers and tusks. *Sci. Today* **9**(2), 37–46.

Dove, W. F. (1935). The physiology of horn growth: A study of the morphogenesis, the interaction of tissues, and the evolutionary processes of a Mendelian recessive character by means of transplantation of tissues. *J. Exp. Zool.* **69**, 347–405.

Dove, W. F. (1936). Artificial production of the fabulous unicorn. A modern interpretation of an ancient myth. *Sci. Mon.* **42**, 431–436.

Dürst, J. -U. (1902). Sur le développement des cornes chez les cavicornes. *Bull. Mus. Hist. Nat. (Paris)* **8**, 197–203.

Fambach (1909). Geweih und Gehorn. Ein kritisches Referat. *Z. Naturwiss.* **81**, 225–264.

Geist, V. (1966a). The evolution of horn-like organs. *Behaviour* **27**, 175–214.

Geist, V. (1966b). Validity of horn segment counts in aging bighorn sheep. *J. Wildl. Manage.* **30**, 634–635.

Grinnell, G. B. (1921). Shed horns of the American antelope. *J. Mammal.* **2**, 116–117.

Hansen, C. G. (1965). Growth and development of desert bighorn sheep. *J. Wildl. Manage.* **29**, 387–391.

Hemming, J. E. (1969). Cemental deposition, tooth succession, and horn development as criteria of age in Dall Sheep. *J. Wildl. Manage.* **33**, 552–558.

Koger, L. M. (1976). Dehorning by injection of calcium chloride. *VM/SAC, Vet. Med. Small Anim. Clin.* **71**, 824–826.

Kômura, T. (1926). Transplantation der Hoerner bei Cavicornien, und Enthornung nach einfachster Methode. *J. Jpn. Soc. Vet. Sci.* **5**, 69–85.

Lankester, E. R. (1907). The origin of the lateral horns of the giraffe in foetal life on the area of the parietal bones. *Proc. Zool. Soc. London* pp. 100–115.

Leister, C. W. (1932). The pronghorn of North America. *Bull. N. Y. Zool. Soc.* **35**(6), 183–193.

Lyon, M. W. (1908). Remarks on the horns and on the systematic position of the American antelope. *Proc. U. S. Natl. Mus.* **34**, 393–402.

MacKinnon, J. (1981). The structure and function of the tusks of babirusa. *Mamm. Rev.* **11,** 37–40.

Marchi, E. (1907). Untersuchungen über die Entwicklung der Hörner bei den Cavicorniern. *Jahrb. Wiss. Prakt. Tierz.* **2,** 32–36.

Murie, A. (1944). "The Wolves of Mount McKinley," Fauna Ser. No. 5, Fauna Natl. Parks U.S., Washington, D.C.

Ness, A. R. (1965). Eruption rates of impeded and unimpeded mandibular incisors of the adult laboratory mouse. *Arch. Oral Biol.* **10,** 439–451.

Noback, C. V. (1932). The deciduous horns of the pronghorn antelope, *Antilocapra americana*. *Bull. N.Y. Zool. Soc.* **35**(6), 195–207.

O'Gara, B. (1969). Horn casting by female pronghorns. *J. Mammal.* **50,** 373–375.

O'Gara, B. W., and Matson, G. (1975). Growth and casting of horns by pronghorns and exfoliation of horns by bovids. *J. Mammal.* **56,** 829–846.

O'Gara, B. W., Moy, R. F., and Bear, G. D. (1971). The annual testicular cycle and horn casting in the pronghorn (*Antilocapra americana*). *J. Mammal.* **52,** 537–544.

Pocock, R. I. (1905). The effects of castration on the horns of a prongbuck (*Antilocapra americana*). *Proc. Zool. Soc. London* pp. 191–197.

Roettcher, D., and Hofmann, R. R. (1970). The ageing of impala from a population in the Kenya Rift valley. *East Afr. Wildl. J.* **8,** 37–42.

Skinner, M. P. (1922). The prong-horn. *J. Mammal.* **3,** 82–105.

Spinage, C. A. (1967). Ageing the Uganda defassa waterbuck *Kobus defassa ugandae* Neumann. *East Afr. Wildl. J.* **5,** 1–17.

Spinage, C. A. (1968). Horns and other bony structures of the skull of the giraffe, and their functional significance. *East Afr. Wildl. J.* **6,** 53–61.

Welles, R. E., and Welles, F. B. (1961). "The Bighorn of Death Valley," Fauna Ser. No. 6. Fauna Natl. Parks U.S., Washington, D.C.

The Evolution of Deer

Over 40 million years ago, before horned or antlered ruminants had evolved, their hornless ancestors roamed the forests and grasslands of the Eocene and Oligocene epochs of both hemispheres. These diminutive ungulates, like some of today's smallest deer, compensated for their lack of horns by the possession of tusklike upper canine teeth. Throughout subsequent evolution in these groups, the development of such tusks has tended to be inversely correlated not only with the size of the animal but with the degree to which horns or antlers were possessed. As the horns grew longer the canines grew shorter, eventually disappearing altogether. In most cervids they have been lost along with the upper incisors.

During the next 10-20 million years, the lower to middle Miocene descendants of these Oligocene ancestors developed a rich variety of outgrowths from their skulls. Although these cephalic appendages may be collectively, albeit inaccurately, referred to as "horns," this does not necessarily imply that they were covered with cornified epidermis. The paleontologist works only with fossilized bones, the composition and surface texture of which do not always make it obvious whether they were enveloped in fur, protected by a horn sheath, or exposed as naked dead bone. During this phase of evolution, there is reason to believe that some of the New World ungulates may have traced their ancestry to Old World forms. However, evidence for migrations in the opposite direction is less convincing.

Fur-Covered Horns

There were at least two major families of extinct ruminants that prevailed from the upper Oligocene to the middle Miocene. The Hypertragulidae and Protoceratidae gave rise to some 4-horned creatures that would have been popular attractions in today's zoos. They culminated in *Syndyoceras* (Fig. 46), which sported a pair of divergent horns from the nasal bone at the end of the snout, balanced by two inward-curving appendages emerging from the skull just posterior to the orbits. Because the skeletal parts of both of these pairs of horns were blunt on their ends, it is believed that they may have been covered in the living animal with a layer of skin and fur, not unlike the horns of today's giraffes. In later related forms, (e.g., *Prosynthetoceras*), the more pointed ends suggest the possibility that bone may have been exposed on their ends (Frick, 1937) as are the bony tips of okapi horns (Lankester, 1907b). These fabulous creatures, whose remains have been unearthed from Nebraska to Texas, became extinct after the middle Miocene. Whether this extinction was brought about by the demise of the

Fig. 46. *Syndyoceras* from the Lower Miocene of Nebraska. (After Frick, 1937.)

last representatives of the group because of climatic changes or unsuccessful competition with more aggressive forms, or by their gradual mutation into different descendants, remains a matter for conjecture.

These ancestors of the giraffe and okapi possessed cranial appendages that were often as elaborate as deer antlers, although it must be assumed that they remained covered throughout life with viable skin. The Giraffidae originated in the Miocene, evolving into animals bearing impressive horns not only in Africa but also in the Pliocene of India (Colbert, 1935; Churcher, 1978). The gigantic *Sivatherium*, shown in Figure 47, carried horns 0.5 m long, horns that were flattened but studded with knobs similar to those sometimes encounted in today's giraffes.

Of special interest are the antlerlike horns of *Climacoceras*, a Miocene giraffoid creature from Africa. *Climacoceras africanus*, discovered near Lake Victoria by MacInnes in 1936, had straight horns over 20 cm long but sprouting up to nine short side branches irregularly arranged along the shaft (Fig. 48). The absence of a burr, together with the continuity of internal vascular channels between horn and skull, confirm that casting and regrowth did not occur. Even more striking are the horns of *Climacoceras gentryi* (Hamilton, 1978a). Growing 30 cm in length, they branched into 5 points, including a brow tine (Fig. 49). Their superficial resemblance to deer antlers is an impressive example of convergent evolution.

Fig. 47. *Sivatherium, a giant giraffid of Africa and India.*

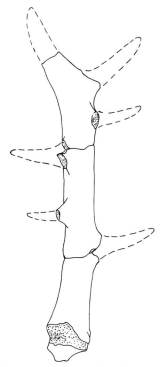

Fig. 48. Fossil horn of *Climacoceras africanus* from the African Miocene. (After MacInnes, 1936.)

However much the horns of giraffes and their relatives may look like antlers, they are independently evolved structures fundamentally different in their modes of development. They originate from "ossicones" in the sub-cutaneous connective tissue (Lankester, 1907a), not from the cranial peri-osteum directly as in the case of deer antlers. The ossicone, which fuses secondarily with the skull, relates the giraffid horn to that of the bovid which is derived from the *os cornu*. They differ primarily in the nature of their integumental coverings.

Giraffid horns are permanent structures with no provision for annual loss and replacement as in the case of antlers. Nor do they shed their skin, except insofar as the distal part of the okapi horn may do so, probably more by accident than design (Lankester, 1907b). The failure of giraffid horns to shed their skin, or themselves to be cast and replaced, is correlated with the tropical habitats in which they evolved.

One of the most prominent families to emerge in the upper Oligocene and flourish in the Miocene was the Palaeomerycidae. The tusked but hornless

Fig. 49. Despite its similarity to an antler, the fossil horn of *Climacoceras gentryi* was a nondeciduous, fur-covered appendage from a Miocene giraffid of Africa. (After Hamilton, 1978a.)

precursors of this family evolved into some of the earliest animals to possess branched headpieces. European deposits have yielded fossils of *Palaeo-meryx* and *Lagomeryx,* diminutive creatures possessing canine tusks as well as horns. The latter sprouted as long slender stems from the frontal bones, surmounted by a crown of several points (Stehlin, 1937). The absence of burrs, or other evidence of possible casting, indicates that these structures were not deciduous. They were probably covered with skin and therefore permanently viable. This characteristic, coupled with the relatively small stature of these animals, suggests that they enjoyed a neotropical climate despite the more temperate conditions that prevail today in the European deposits where their fossils have been found. It is doubtful that appendages such as theirs could have escaped frostbite if exposed to wintry climates.

Nondeciduous "Antlers"

Deer antlers not only shed their velvet, but also drop off and are replaced each year. However, it is of interest to note that deer were not the first to develop shedding of the skin. During the Miocene epoch in North America, the Merycodontidae grew branched appendages that lost their skin but remained permanently thereafter as dead bony structures. Although these animals were antecedents of the Antilocapridae, not the Cervidae, their horns superficially resembled antlers despite some very basic differences that put them in a class by themselves. It is to Childs Frick (1937) that we owe much of our knowledge of these interesting animals.

Cosoryx, first described in 1869, is represented by numerous Miocene specimens from the rich fossiliferous deposits of Nebraska, Kansas, Colorado, Montana, and New Mexico. This small deerlike creature grew bifurcate horns that often possessed one to several annular flanges or "pseudoburrs," around the proximal portions of their pedicles (Fig. 50). Otherwise, the

Fig. 50. *Paracosoryx serpentinus*, a fossil merycodont from the North American Miocene. (Frick Collection, American Museum of Natural History.)

surface texture of the *Cosoryx* horn was relatively smooth. As early as 1890, Scott and Osborn concluded that this horn was "almost certainly covered with skin; its smooth surface . . . shows that it could not have been naked, as in the true deer." Equally interesting horns occurred in other forms. In some, such as *Ramoceros*, up to 4 points grew (Fig. 51). In *Meryceros* (Figs. 52, 53) and *Merycodus*, 2–4 points were typically present, as were the curious pseudoburrs so reminiscent of the basal burrs present in today's antlers.

If the merycodontid "horns" were in fact true antlers, which by definition are lost and replaced from time to time, it is difficult to explain why fossils that had cast them are not found (Furlong, 1927). If these structures were replaced annually, one would expect to find specimens of cast antlers, skulls from which they had been lost, and an age series of progressively larger antlers with increasing number of points. No such fossils have been

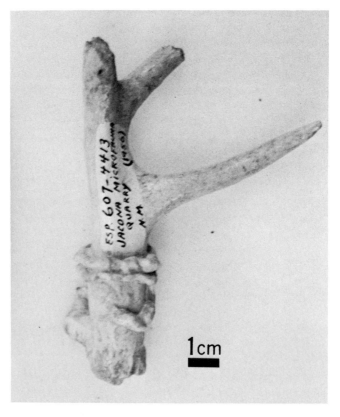

Fig. 51. Nondeciduous antler of *Ramoceros*, showing three pseudoburrs, the most basal of which is quite irregular. (Frick Collection, American Museum of Natural History.)

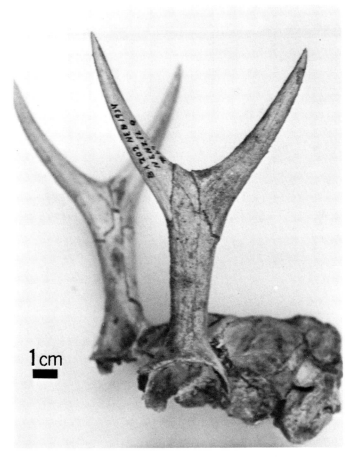

Fig. 52. *Meryceros nenzilensis,* presumably a young specimen with nondeciduous antlers still enveloped in viable skin and lacking pseudoburrs. (Frick Collection, American Museum of Natural History.)

discovered. It has been suggested that the absence of detached antlers could be accounted for if these animals had migrated seasonally to the highlands during their antler-growing times of year (Matthew, 1904). At higher elevations, subsequent erosion would have wiped out fossil remains, while conditions more favorable to the preservation of their remains might have prevailed in the flatlands where such animals could have spent the colder times of year.

The existence of more than one pseudoburr (Figs. 51, 54) around the basal portions of such horns is a condition that is never encountered in modern deer. The burrs of existing antlers are cast along with the antler

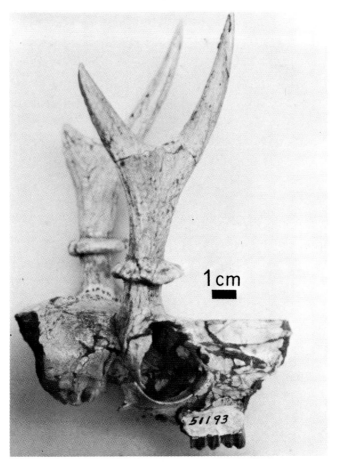

Fig. 53. *Meryceros crucensis* illustrating single pseudoburrs on its nondeciduous antlers. (Frick Collection, American Museum of Natural History.)

itself. If the merycodontid antlers were lost annually, each set would have had to separate from the pedicle above the burr, not below it, and each year a new burr would be expected to form distal to the previous ones. Although such a sequence of events would not be impossible, there is little or no reason to conclude that the pseudoburrs were in fact associated with the casting of these outgrowths. There may well be a more logical explanation for their existence.

Equally untenable is the notion that these outgrowths represent bony horn cores. Yet Pilgrim (1941) suggested that pseudoburrs might have developed

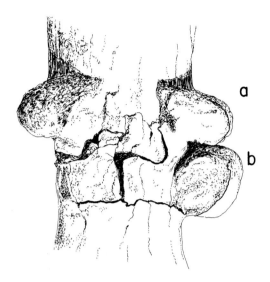

Fig. 54. *Merycodus* pseudoburrs, showing their relation to underlying bone where fragments have been broken away. The more distal pseudoburr (a) is continuous with a bony lamina that is overlain by the one from which the more proximal pseudoburr was developed (b). This is evidence that the pseudoburrs were formed in a distoproximal time sequence. (After Voorhies, 1969.)

in association with the annual shedding and regrowth of horn sheaths. For obvious reasons, such horn sheaths could not have been shed in the same manner that present day pronghorn sheaths are. To achieve such a feat, it would be necessary to possess a tapered bone core from which the sheath could easily slip off when the time comes. The only way such a horn sheath could be lost from branched bone cores would be by piecemeal cracking or flaking. Matthew (1924) suggested that they could have been "covered during life with a velvet or with a horny casing which split or peeled off periodically but not until a new protective casing had formed within it, the burr then representing the seasonal stoppage of growth and lime supply after the new growth of horn and horn-case had been completed." However, it would be virtually impossible to grow hardened horn sheaths on obtusely branched bone cores. Unlike antlers, horns grow at their bases, pushing the previously formed horn sheaths distally as new material is laid down proximally. The mechanics of this process dictate that the inside dimensions of the distal parts cannot be greater in diameter than the proximal portions. Therefore, it would be impossible to produce successive increments of cornified tissue on the surface of a branched horn core unless each branch were to constitute a separate horn sheath, or unless adjacent ones did not diverge (Webb, 1973).

For the aforementioned reasons, it must be concluded that the merycodontid horns were neither antlers, subject to periodic casting, nor cornified horns. "The actual method of growth and nature of the covering of the Merycodontini 'horn' remains a conundrum," wrote Frick in 1937, "though the available evidence as here interpreted points to the probability of there having been, in such forms as *Ramoceros*, a from time to time replacement somewhat after the method in the Recent deer." Frick went on to note that these structures were composed of compact bone peripherally and more porous bone internally. This would be consistent with the possibility that they were dead exposed bony appendages in the living animal, at least after their full growth had been attained. Furlong (1927) believed that "the burrs, then, may be regarded as the terminal point of the heavier skin covering of the head, and the initial point of growth of the lighter covering of the antlers." Frick (1937) agreed that "multiple burrs might be accounted for through periodical retreat of the velvet and encroachment of the head covering." If this is taken to indicate that new skin might grow over dead bone, it would be a phenomenon without precedent in other biological systems. Nevertheless, the merycodont horn would appear to have shed its skin to yield a dead bony appendage but was itself neither cast nor replaced from time to time.

This is supported by the fact that, unlike the antlers of today's deer, the fossil record does not yield a series of progressively complex structures as might be expected to have been produced during the first few years of life while the animal was maturing. Indeed, the apparent absence of unbranched horns is particularly significant. Only in the most immature specimens are short, unbranched horns found, structures that were clearly in the early stages of their original development at the time of death (Voorhies, 1969). Thus, if these were replaceable appendages, they evidently did not become increasingly complex with age, suggesting that merycodonts may have attained their full stature relatively early in life. For unaccountable reasons, there is considerable individual variation in the shapes and sizes of these horns from one specimen to another.

Close examination of the pseudoburrs reveals that they tend not to be so much an outgrowth from the underlying bone as a periodic deposition of bone on the surface. In this respect they differ from the true burrs of antlers. In some specimens they have been shown to be relatively unattached to the subjacent bone, often easily broken off (Furlong, 1927). It is also worth noting, as Voorhies (1969) illustrated (Fig. 54), that each of the pseudoburrs is associated with a separate outer layer of bone. This situation would be consistent with the possibility that the pedicle might increase in diameter each year by the addition of circumferential lamellae on the surface of existing bone. Such an interpretation of the facts supports the contention of Furlong (1927) and Frick (1937) that the skin of the horn might have regressed proximally each year, the bony pseudoburr representing successive levels to which the viable skin of the pedicle receded.

According to this scheme, the original horns grew out to their full dimensions as skin-covered appendages. Because these horns occurred only in

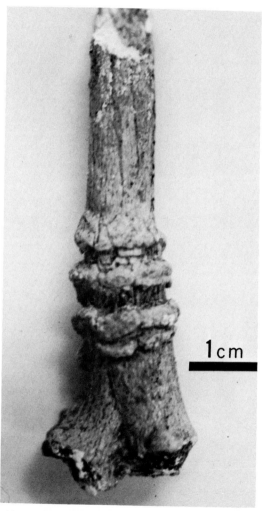

Fig. 55. Nondeciduous antler of *Paracosoryx wilsoni* with four pseudoburrs. (Frick Collection, American Museum of Natural History.)

males, they might well have been subject to repeated injuries of combat. Alternatively, despite the relatively warm climate of the Miocene, the horns would still have been vulnerable to freezing in the winter at higher latitudes. It is also conceivable that shedding of the skin could have been hormonally induced, as in deer antlers. In any case, dead bony protuberances would have resulted. These fossils lacking pseudoburrs presumably died while their horns were still viable (Fig. 52). Those with a single pseudoburr would have survived until the horns died and their skin covering was lost as far proximally as indicated by the burr (Figs. 50, 53). Those with two or more pseudoburrs probably survived for several more years, losing a distal ring of pedicle skin each winter, the remnants of which would have laid down successively more proximal bony annuli (Figs. 51, 55) that resemble only superficially the burrs of deer antlers. Indeed, the somewhat irregular and often incomplete nature of these pseudoburrs suggests the uneven pattern of integumentary regression that might have been expected in the above circumstances.

The existence of bony outgrowths from the skull that shed their skin but are not themselves cast strains the definition of antlers. In the absence of a cornified sheath, they cannot logically be referred to as horns. Alternatively, they resemble the horn core, that is, the bony superstructure upon which horn sheaths are deposited. Yet no extant horn core is naturally denuded and allowed to die as were the merycodontid ones. As such, these were unique structures more closely resembling antlers than horns despite their failure to be cast each year. It is therefore proposed, after Whitworth (1958), that they be referred to as "nondeciduous antlers" in recognition of the fact that they lost their velvet, died, but remained permanently attached to the skulls of their possessors.

The Earliest Antlers

From the foregoing account, it is clear that during the early middle Miocene there was no dearth of potential ancestors of the Cervidae. It is entirely possible that none of the known forms was in fact the actual precursor of modern deer. However, these fossils indicate that a rich diversity of cephalic appendages evolved in various forms at different times and locations, often independently of each other. Although merycodonts were not related to deer, their horns were probably the closest approximation to antlers as they are presently known. They represented dead bony appendages from which the covering skin, either by accident or design, was shed. However, they did not take the next evolutionary step of actively casting the dead bony parts, or of regrowing replacements.

The world's earliest known deer were *Dicrocerus* and *Stephanocemas*. Fossils of both have been discovered in Miocene deposits of Europe and Asia, where they survived into the Pliocene epoch. Both had long pedicles surmounted by antlers that were unmistakably cast, judging from the occurrence of burrs at their bases, as well as the existence of cast-off specimens.

Dicrocerus was a small deer with antlers that typically had 2 points (Fig. 56). Over a century ago, Dawkins (1878) described them as "springing close to the burr, and crowning the summit of a long and slender pedicle like that of the Muntjak." Stehlin (1939) figures unbranched specimens, presumably from yearlings, as well as an antler with 3 points; simple forked outgrowths were clearly the rule, however. In older individuals the base of the antler tended to be wider than the pedicle supporting it (Stehlin, 1939).

Stephanocemas had more elaborate antlers (as well as tusklike upper canines). The pedicles were sometimes 12 cm long. Perched on top like a cup was a palmate antler unlike anything seen in modern living deer (Fig. 57). Up to 7 points radiated out from the flattened central portion, but considerable variation is seen from one specimen to another. When ar-

Fig. 56. *Dicrocerus,* one of the earliest cervids to evolve, carried true antlers that were cast and replaced each year. (After Stehlin, 1939.)

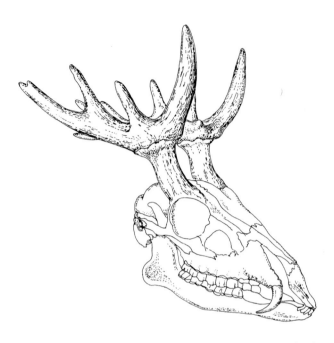

Fig. 57. *Stephanocemas* was a small deer from the Miocene of Mongolia. Its antlers developed increasing numbers of points as they were replaced in successive years. (After Colbert, 1936.)

ranged in order of increasing size and number of points, they have been interpreted by Colbert (1936) as representing successively greater ages of the deer from which they came. The deciduous nature of *Stephanocemas* antlers is further indicated by the presence of a burr at the top of the pedicle, and the existence of cast specimens. However, burrs were not conspicuously developed in these forms.

   If the first cervids, as defined by the possession of deciduous antlers, evolved in the Miocene of Asia and Europe, it is important to know what the climates of these regions might have been at that period in the earth's history. Although the deposits where *Stephanocemas* was first unearthed are in Mongolia at a latitude that is presently in the temperate zone, there is reason to suspect that the climate might have once been at least neotropical. Other fossils from such regions (e.g., elephants) were of animals currently associated with warmer climates. Moreover, the Himalayan Mountains did not originate until the process of continental drift brought what is now the Indian peninsula in contact with the Asiatic mainland, a geologic development that did not occur until the Miocene. Probably tropical climates prevailed much farther north into China before such moderating influences were cut off by the Himalayas. If this reasoning is correct, the nondeciduous precursors of modern antlers may have in fact evolved in warmer climates, acquiring the capacity for self-replacement only when the animals migrated to temperate zones, or were overtaken by gradual climatic cooling in their native habitats.

The descendants of these earliest deer radiated from their presumed center of origin in Asia to Europe and North America. Not until the lower Pliocene did they make their way to India, presumably delayed by the Himalayan barrier. Although they populated the East Indies, they were no more successful in reaching Australia than were most other mammals.

The World's Largest Deer

Of all the fossil deer that have been unearthed, none has excited man's imagination as much as the various gigantic specimens that have been discovered in the Pleistocene of both hemispheres. As Geist (1974) has emphasized, this was a period of accelerated evolution in which many bizarre giants were produced in various mammalian families. From the late Pleistocene of North America, a giant deer-moose, *Cervalces,* was found in New Jersey by Scott (1885). Its massive antlers, measuring 1.5 m from tip to tip, grew laterally from the skull in much the same way the antlers of modern moose do. *Cervalces* had two palmations on each antler, the main one projecting horizontally to the side, and another growing forward in a scoop-shaped configuration (Fig. 58). Because of the latter's location on either side of the head, they would have blocked the lateral vision of the animal. It is assumed that the inconvenience of these blinders was compensated by the impressions they must have made on other members of the species.

In Villafranchian deposits of the early Pleistocene in Europe, a so-called "bush-antlered deer," *Eucladoceros (Euctenoceros),* has been found. A specimen from northern Italy carried antlers measuring over 1 m in length. These enormous antlers branched and rebranched into as many as 12 points each (Rütimeyer, 1881, 1883), ranking them among the most complex antlers ever to have evolved (Fig. 59).

Far more impressive than either *Cervalces* or *Eucladoceros* was *Megaceros giganteus,* or the so-called Irish elk, hundreds of skulls of which have been dug from beneath the peat bogs of Ireland. The Irish elk is not an elk, nor is it necessarily Irish, its remains having been found on the European continent and as far east as southern Japan. Nevertheless, the British Isles, and in particular Ireland, is where this species flourished, or at least where the conditions necessary for its fossilization have been most favorable.

Fig. 58.   Reconstruction of *Cervalces,* based on a fossil illustrated by Scott (1885).

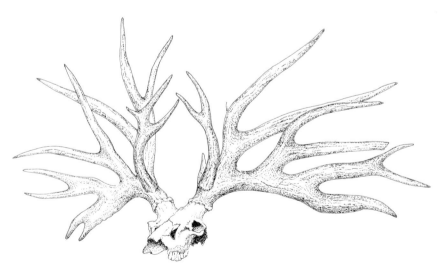

Fig. 59.   The bush-antlered deer, *Eucladoceros dicranios,* from the Pleistocene of Europe. (After Rütimeyer, 1881.)

The antlers of *Megaceros giganteus* weighed up to 45 kg, and perhaps more when they were in velvet (Coope, 1973). These massive antlers, whose weights were over ten times that of the skull from which they grew, projected laterally and horizontally as magnificently palmated headpieces (Fig. 60). They also possessed brow tines that grew forward from the base of the main beam, sometimes as horizontal shovel-shaped projections (Fig. 61) like visors over the eyes (Kahlke, 1955). The exaggerated size of these antlers, coupled with the fact that they usually did not grow vertically or forward, suggests that they were more useful for display than combat. Although these were the largest antlers ever grown by deer, *Megaceros* itself was not necessarily larger than the modern moose (Coope, 1973). For his size, the moose often carries relatively small antlers notwithstanding their impressive dimensions in absolute terms. However, in the case of *Megaceros*, the disproportionate size of the antlers is consistent with Huxley's (1932) proposal that deer antlers represent a case of positive allometry. The allometric equation

$$x = by^k$$

states that the dimension of an organ (x) is a fraction (b) of the body weight (y) raised to a given power (k), the latter being $k > 1$ or $k < 1$ depending on

Fig. 60. Fossil skeleton of *Megaceros giganteus* on exhibit at the British Museum of Natural History.

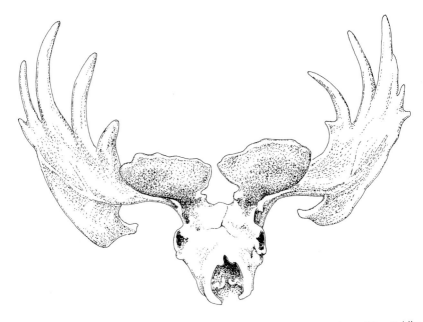

Fig. 61. *Megaceros* specimen displaying horizontally flattened brow tines. (After Kahlke, 1955.)

whether the body part grows faster or slower than the rest of the organism. If various sized deer species are compared, it is noted that the smallest ones have antlers that are comparatively underdeveloped, although the largest have extra big antlers. In other words, when the sizes of animals change, the dimensions of their antlers change even more. *Megaceros* is no exception to this principle, as Gould (1973) has shown by comparing numerous skulls from specimens of different sizes. Record antlers of *Megaceros* have grown to lengths of 174 cm, with a span of over 3 m from tip to tip. Not surprisingly, their neck vertebrae were also enlarged. How these animals could have afforded the luxury of growing such extravagant antlers, the mineral components of which comprised a significant fraction of the body's entire skeletal material, has been an intriguing puzzle to paleontologists. Indeed, it has been claimed that the eventual extinction of *Megaceros* might have been caused by the disadvantages of its positive allometry to the extent that its antlers became a luxury it could neither afford nor get rid of.

If it is assumed that *Megaceros* replaced his antlers yearly, and that the new ones grew during a span of 3–4 months as is the case with all extant deer, then the rate of elongation must have exceeded several centimeters per day at the steepest inflection of its growth curve in the late spring or early summer. It is perhaps significant that unbranched spike antlers of Megaceros are rarely found (Thenius, 1958). In most lesser species of deer, yearlings typically grow unbranched antlers. In Megaceros, the deer were so large that even the first set of antlers may have tended to be branched.

The mobilization of sufficient materials for producing such gigantic antlers must have severely taxed the metabolism of these deer. The massive amounts of calcium needed to deposit so much bone in such a short amount of time would have demanded an unusually high calcium content in the deer's diet. The calcium-rich deposits in which these fossils have been found in Ireland suggest that such conditions must have prevailed there (Mitchell and Parkes, 1949). They also raise the possibility that if climatic and geologic changes brought about a decrease in the availability of calcium, the survival of Megaceros might have been seriously threatened.

Megaceros flourished in the late Pleistocene, becoming extinct about 10,000 years ago in Ireland, but possibly surviving until 2,500 years ago in the Black Sea region (Kurtén, 1968). Its remains were first discovered in Ireland in 1588, and in following centuries it was considered something of a status symbol to find one's own fossil for display in the mansions and castles of the British Isles (Mitchell and Parkes, 1949). Skulls were most commonly located in the silty clay, or shell-marl, beneath the peat deposits of Ireland. Fossils have been found anywhere up to 15 m below the surface, sometimes incidental to excavations for other purposes, but on occasions their discovery was no accident. By probing through the layers of peat with a long metal rod, an experienced fossil-hunter could not only locate solid objects at some depth, but could even distinguish bones from stones (Millais, 1897).

Judging from the locations of such skulls, Megaceros inhabited the plains and lowlands rather than the higher elevations, presumably because they preferred grazing to browsing. Curiously, most of the skulls that have been unearthed have been found upside down, probably because the center of gravity was in the antlers rather than the skull itself. Since submerged carcasses were presumably the only ones that became fossilized, it is logical that the head would have tended to become lodged upside down in the mud at the bottom of ponds and streams.

Megaceros coexisted in Europe with ancestors of the hippopotamus, rhinoceros, and musk ox (Millais, 1897). Whether the giant deer was contemporaneous with early man is not known, and may have depended on the geographic location. If Megaceros survived in the Black Sea region into historic times, as Kurtén (1968) has suggested, there is a good possibility that it coexisted with early man in that region. In Ireland, there is no unequivocal evidence for contemporaneity. Although Megaceros bones have been found in Irish caves, they are always associated with the bones of such predators as bears and hyenas which were probably responsible for leaving them there

(Mitchell and Parkes, 1949). Sometimes the antlers show evidence of scratches, as if they may have been used as a cutting board. However, it is equally conceivable that the scratches were made as the antlers were carried downstream to their final resting places. Millais (1906) believed that the ancestors of the Lapps may have hunted giant deer in Scotland. He also described a *Megaceros* skull "on the forehead of which there is a deep cut evidently made by a stone hatchet." The evidence is circumstantial.

Perhaps the most convincing evidence that *Megaceros* became extinct before the advent of man is that illustrations of these giant deer are conspicuous by their near absence in the paintings left in European caves by Palaeolithic man. It is unlikely that our ancestors would not have drawn these magnificent creatures, had they known them, along with the bison, mammoths, red deer, and reindeer with which they embellished the walls of their dwellings. Yet only in Pech Merl (Breuil, 1952) and Cougnac (Leroi-Gourhan, 1965) caves in France have wall paintings of *Megaceros* been discovered (Fig. 12). The absence of roe deer illustrations, which must have flourished in prehistoric Europe, is equally inexplicable. As tempting as it is to imagine that such an ostentatious beast as *Megaceros* may have been exterminated by Pleistocene man, it is likely that more global influences were responsible for its extinction. The advance of ice sheets during the last glacial period abolished many of the habitats previously occupied by *Megaceros* (Coope, 1973). Unadapted to survival on tundra vegetation, these animals apparently disappeared with the flora on which they depended, eventually to be replaced by animals such as the reindeer that were more adapted to the new conditions. Even if nature had not ensured their extinction in the Pleistocene, it is difficult to imagine how these irresistible game animals could have survived the predatory instincts of civilized man.

Continent without Deer

With only a few exceptions (Hamilton, 1978b), deer have failed to invade the African continent. The Barbary stag (*Cervus elaphus barbarus*), a subspecies of the red deer, inhabits parts of northwest Africa. This deer is believed to have immigrated by way of Sicily during the Pleistocene epoch. The giant deer (*Megaceros algericus*) followed the same route, leaving its fossil remains in Algeria until the early Holocene. The fallow deer invaded Egypt by way of Suez in the late Pliocene. It managed to spread as far south as Ethiopia before progressive desiccation of its habitats contributed to its extinction in Africa during the Pleistocene.

It is a curious fact that tropical Africa is currently as poor in deer as it is rich in horned forms, including giraffids. Even if the Sahara desert were not always the barrier it is today, one wonders if the ecological competition between deer and antelopes might not have discouraged their coexistence. A reciprocal situation occurs in South America where horned ungulates are lacking but various species of deer abound. Likewise, in southeast Asia there are many species of deer to the relative exclusion of horned animals. On the Indian subcontinent horned and antlered animals coexist, as they do in north temperate latitudes.

Whether or not Africa has always been as devoid of deer as it is today is open to question. Joleaud (1935) suggested possible importations of deer by the Phoenicians. Keimer (1934) described two fragments of fallow deer antlers from Thebes, comparing these with deer illustrated in ancient Egyptian art (Fig. 62). Dawson (1934) also confirmed the Egyptians' knowledge of deer, although the graphic talents with which they were drawn did not always make it obvious what species was involved. It has been suggested that such apparent inaccuracies might have represented the stylized impressions of relatively unfamiliar animals. The nearest species to Egypt would have been the Mesopotamian fallow deer, stories of which might have been distorted in the retelling. The Egyptian deer were sometimes spotted, and in profile revealed the prominent penis so characteristic of fallow males.

The antlers on the deer depicted in ancient Egyptian art typically possessed relatively straight beams with side branches sprouting at irregular intervals along their lengths. Such a configuration is totally unlike the antlers of fallow deer which tend to be palmate and in any case often exhibit

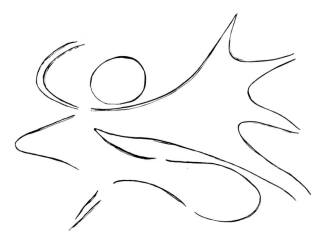

Fig. 62. Stylized illustration of *Megaceros* in Pech Merl cave in France, suggesting that giant deer may have coexisted with early man on the European continent. (After Breuil, 1952.)

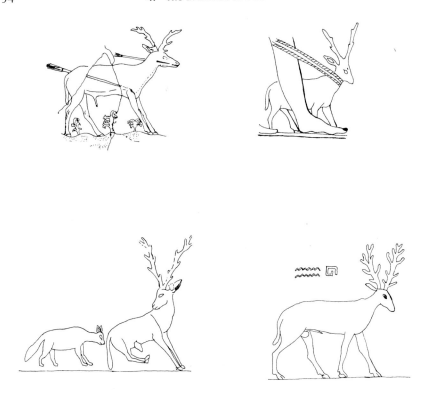

Fig. 63.   Examples of deer in ancient Egyptian art. Despite the absence of palmate antlers, the relatively long tail and penis suggest that they were fallow deer. (After Keimer, 1934.)

dichotomous branching. MacInnes (1936) noted the striking similarity between the horns of *Climacoceras africanus,* a Miocene giraffoid animal he believed to be a deer, and the more or less conventionalized antlers illustrated on the monuments, tombs, and wall paintings of ancient Egypt (cf. Figs. 48 and 63). He speculated that this was more than just a curious coincidence, even hinting that *Climacoceras* might have survived to recent times in some isolated region of central Africa!

Deer in the Americas

The history of the North American deer is more speculated about than understood. The only certainty is that wapiti descended from Asiatic immigrants that left their fossil remains in Alaska during interglacial times

(Guthrie, 1966). The capacity for wapiti and red deer to interbreed testifies to their relatedness.

The fossil record is not adequate to establish how New World deer may be related to those of the Old World. There is a remote possibility that they could have arisen independently, but the probability that replaceable antlers might have evolved more than once is unacceptable, especially in view of the absence of fossil deer prior to the Pliocene epoch in North America.

Deer in the two hemispheres differ in two major respects (Chapter 2). New World deer are classified in the Telemetacarpalia, that is, their second and fifth metacarpals are present only as distal skeletal remnants associated with the dewclaws (Fig. 28). In most Old World deer, proximal segments of these metacarpals also persist, this being diagnostic of the Plesiometacarpalia (Brooke, 1878).

The other distinguishing characteristic between New World and Old World deer is their chromosome numbers (Table I). Most North and South American deer possess 70 chromosomes, although the vast majority of Old World deer do not. Although there are interesting exceptions that undermine the taxonomic value of karyotypes, chromosome numbers can be useful in association with other characteristics as indicators of how animals may be related to each other.

It is significant that the world's only circumpolar species of deer both possess 70 chromosomes and are in the Telemetacarpalia. In these features, therefore, moose and reindeer may be related to New World deer. Fossil evidence suggests that the moose probably arose in Eurasia, but from unknown ancestors. Reindeer and caribou, which evolved relatively recently during the Ice Age, are believed to have originated in the New World (Flerov, 1952; Frick, 1937).

This still leaves unexplained the derivation of *Odocoileus* and its relatives. Its earliest fossils are found in Pliocene deposits (Fry and Gustafson, 1974), and these give no clue to its earlier ancestry. The lack of factual evidence on the origins of *Odocoileus* has promoted considerable speculation. Cameron (1892b) suggested that the white-tailed deer may have arisen independently in the middle Miocene of North America, before populating South America. Matthew (1908) mentioned the possibility that it could have evolved from *Blastomeryx*, a hornless inhabitant of Miocene America. Hershkovitz (1969) claimed that it may have evolved in Central America as recently as the Pleistocene, following the appearance of various South American genera. Probably none of these explanations is correct.

It is more likely that *Odocoileus* probably descended from an Old World ancestor. If the moose and reindeer are ruled out as the forerunner of other North and South American deer, this leaves the roe deer and the Chinese water deer as the only other members of the Telemetacarpalia with 70 chromosomes. The roe deer would be the more likely progenitor of *Odocoileus*, a possibility alluded to by Millais in 1906. They both live in northerly latitudes in their respective hemispheres and occupy similar ecologic niches. Coincidentally, both deer lack typical brow tines on their antlers, if this counts for anything. In the roe deer, the antlers grow straight up from the skull for some distance before branching. In the white-tail, what is sometimes taken for the brow tine is frequently referred to as a "basal snag" because it arises not on the front of the shaft but projects vertically from its median surface.

Despite the numerous differences between *Capreolus* and *Odocoileus*, the foregoing characters shared by them lend some support to the hypothesis of a common ancestor. If such a hypothetical precursor had migrated to America in the Pliocene only to be cut off from its Eurasian cousins, the two populations might have diverged sufficiently in the course of subsequent evolution to have become superficially different. The evidence that this actually happened is too circumstantial to be taken seriously, yet it is compelling enough to warrant the continued search for facts that may help to explain why the ranges of these two deer do not overlap.

Whatever and wherever may have been the origins of *Odocoileus*, this interesting deer has been eminently successful in extending its habitat over a greater range of latitudes than any other cervid genus. Even before the Panamanian isthmus was established a few million years ago, there is reason to believe that white-tailed deer may have island-hopped their way to South America (Hershkovitz, 1969). It currently lives north of the Amazon and along the western coast as far south as Peru. The striking similarities between the white-tailed deer and the pampas and swamp deer, if not the huemul, suggest the possibility of a common ancestor. Even the tiny pudu must be related. In addition to having 70 chromosomes, it is, like the pampas deer and the huemul, in the Telemetacarpalia group. Whether or not the brockets share this ancestry remains to be proven. Nevertheless, it is difficult to imagine that all South American deer did not have a common origin presumably involving *Odocoileus* or its forerunner (Kraglievich, 1929–1932). Thus, the major unanswered question remains the derivation of *Odocoileus* itself, and the solution to this challenging problem must await the discovery of more meaningful fossils that will shed light on how the deer of the New World relate to those of the Old World.

Recapitulation

It was almost inevitable that students of cervid evolution should not have failed to note the similarities between the phylogenetic and ontogenetic development of antlers. From *Dicrocerus* in the middle Miocene to the bush-antlered deer in the Pleistocene and the highly branched antlers of many modern deer, there has clearly been a general increase in the number of points throughout the geologic history of the Cervidae. In 1881, Dawkins wrote that "the development of antlers indicated at successive and widely separated pages of the geological record is the same as that observed in the history of a single living species." Earlier (Dawkins, 1878), he had stated that "these successive changes are analogous to those which are to be observed in the development of antlers in the living deer, which begin with a simple point and increase their number of tynes until their limit be reached." Cameron (1892a) echoed the same belief:

> There is no trace of cranial weapons in the fossils of the Lower Miocene, where the absence of antlers is characteristic of the infant race, as now of the infant deer. . . . It becomes apparent that the period embracing calfhood, first antlers, and second antlers in the growth of living deer corresponds to the Miocene chapter of their family record, that with the third antlers we pass into the Pliocene and onwards.

In more recent years, Schindewolf (1946) and Geist (1971) have also pointed out these progressions from the simple to the complex.

The question is not whether the ontogeny of antlers resembles their phylogeny, which it does, but what the meaning of this apparent recapitulation might be. Despite the superficial similarities between antler evolution and antler growth, there is no reason to conclude that development is retracing geologic history in the maturation of each deer. It would seem more logical to explain the increasing complexities of antlers, whether phylogenetic or ontogenetic, on the basis of increasing body sizes. In his allometric studies of deer antlers, Huxley (1926, 1931) calculated a positive growth coefficient of about 1.5 for red deer during the first one-half of life, meaning that antler size increases one-half again as much as the body during the course of maturation. In general, progressively larger deer, whether they be young to old or small species to large ones, grow disproportionately larger antlers. Gould's (1973) measurements on the Irish elk confirm a growth coefficient as high as 2.5 in this species. However, in the roe deer, it is only 0.57 (Huxley, 1931).

The implications of this are interesting. As small ancestral deer evolved into larger descendants, antler size and complexity increased. A comparable phenomenon is observed when yearlings are compared with fully grown adults. In both cases, it is possible that antler size may simply be a function of pedicle diameter. Bigger antler buds grow into longer antlers, which in turn produce more points. This raises the intriguing question of what the upper limits of antler size and complexity might be. Are the morphologic characteristics of antlers determined by genetic potentials, or are they a straightforward derivative of pedicle size? To answer this question it would be necessary to enlarge the pedicle artifically to see if it would give rise to a super-antler.

References

Breuil, H. (1952). "Quatre cents siècles d'art parietal." Montignac, Dordogne, France.

Brooke, V. (1878). On the classification of the Cervidae, with a synopsis of the existing species. J. Zool. pp. 883–928.

Cameron, A. G. (1892a). Value of the antlers in classification of deer. Field **2055**, 703.

Cameron, A. G. (1892b). Value of the antlers in the classification of deer. New World antlers of Pliocene divergence. Field **2059**, 860–861.

Churcher, C. S. (1978). Giraffidae. In "Evolution of African Mammals" (V. J. Maglie and H. B. S. Cooke, eds.), pp. 509–535. Harvard Univ. Press, Cambridge, Massachusetts.

Colbert, E. H. (1935). Siwalik mammals in the American Museum of Natural History. Trans. Am. Philos. Soc. [N. S.] **26**, 1–401.

Colbert, E. H. (1936). Tertiary deer discovered by the American Museum Asiatic expeditions. Am. Mus. Novit. **854**, 1–21.

Coope, G. R. (1973). The ancient world of "Megaceros." Deer **2**, 974–977.

Dawkins, W. B. (1878). Contributions to the history of the deer of the European Miocene and Pliocene strata. Q. J. Geol. Soc. London **34**, 402–420.

Dawkins, W. B. (1881). On the evolution of antlers in the ruminants. Nature (London) **25**, 84–86.

Dawson, W. R. (1934). Deer in ancient Egypt. J. Linn. Soc. London, Zool. **39**, 137–145.

Flerov, K. K. (1952). Musk deer and deer. In "Fauna of USSR. Mammals". Acad. Sci. USSR. Moscow.

Frick, C. (1937). Horned ruminants of North America. Bull. Am. Mus. Nat. Hist. **69**, 1–669.

Fry, W. E., and Gustafson, E. P. (1974). Cervids from the Pliocene and Pleistocene of central Washington. J. Paleontol. **48**, 375–386.

Furlong, E. L. (1927). The occurrence and phylogenetic status of Merycodus from the Mohave Desert Tertiary. Univ. Calif., Berkeley, Publ. Geol. Sci. **17**, 145–186.

Geist, V. (1971). The relation of social evolution and dispersal in ungulates during the Pleistocene with emphasis on the Old World deer and the genus Bison. Quat. Res. (N.Y.) **1**, 285–315.

Geist, V. (1974). On the relationship of ecology and behaviour in the evolution of ungulates: Theoretical considerations. In "The Behaviour of Ungulates and its Relation to Management" (V. Geist and F. Walther, eds.), New Ser. Publ. 24, Vol. I, pp. 235–246. IUCN, Morges, Switzerland.

Gould, S. J. (1973). The misnamed, mistreated, and misunderstood Irish elk. Nat. Hist., N.Y. **82**(3), 10–19.

Guthrie, R. D. (1966). The extinct wapiti of Alaska and Yukon Territory. Can. J. Zool. **44**, 47–57.

Hamilton, W. R. (1978a). Fossil giraffes from the Miocene of Africa and a revision of the phylogeny of the Giraffoidea. Philos. Trans. R. Soc. London, Ser. B **283**, 165–229.

Hamilton, W. R. (1978b). Cervidae and Palaeomerycidae. In "Evolution of African Mammals" (V. J. Maglio and H. B. S. Cooke, eds.), pp. 496–508. Harvard Univ. Press, Cambridge, Massachusetts.

Hershkovitz, P. (1969). The evolution of mammals on southern continents. VI. The recent mammals of the Neotropical Region: A zoogeographic and ecological review. *Q. Rev. Biol.* **44**, 1–70.

Huxley, J. S. (1926). The annual increment of the antlers of the red deer (*Cervus elaphus*). *Proc. Zool. Soc. London* **67**, 1021–1036.

Huxley, J. S. (1931). The relative size of antlers in deer. *Proc. Zool. Soc. London* pp. 819–864.

Huxley, J. S. (1932). "Problems of Relative Growth." Methuen, London.

Joleaud, L. (1935). Les ruminants cervicornes d'Afrique. *Mem. Inst. Egypte* **27**, 1–85.

Kahlke, H. D. (1955). "Grossäugetiere im Eiszeitalter." Urania-Verlag, Leipzig.

Keimer, L. (1934). Sur deux fragments de cornes de daim trouvés à Deir el-Médineh. *Mélanges Maspero. I. Orient Ancien* **1**, 273–308.

Kraglievich, L. (1929–1932). Ciervos fosiles del Uruguay. *An. Mus. Hist. Nat. Montevideo* [2] **3**, 355–435.

Kurtén, B. (1968). "Pleistocene Mammals of Europe." Aldine Publ., Chicago.

Lankester, E. R. (1907a). The origin of the lateral horns of the giraffe in foetal life on the area of the parietal bones. *Proc. Zool. Soc. London* pp. 100–115.

Lankester, E. R. (1907b). On the existence of rudimentary antlers in the okapi. *Proc. Zool. Soc. London* pp. 126–134.

Leroi-Gourhan, A. (1965). "Préhistoire de l'art occidental." Éditions d'Art Lucien Mazenod, Paris.

MacInnes, D. G. (1936). A new genus of fossil deer from the Miocene of Africa. *J. Linn. Soc. London, Zool.* **39**, 521–530.

Matthew, W. D. (1904). A complete skeleton of *Merycodus*. *Bull. Am. Mus. Nat. Hist.* **20**, 101–129.

Matthew, W. D. (1908). Osteology of *Blastomeryx* and phylogeny of the American Cervidae. *Bull. Am. Mus. ,Nat. Hist.* **24**, 535–562.

Matthew, W. D. (1924). Third contribution to the Snake Creek fauna. *Bull. Am. Mus. Nat. Hist.* **50**, 59–210.

Millais, J. G. (1897). "British Deer and Their Horns." Henry Sotheran & Co., London.

Millais, J. G. (1906). "The Mammals of Great Britain and Ireland," Vol. 3, pp. 71–179. Longmans, Green, London.

Mitchell, G. F., and Parkes, H. M. (1949). The giant deer in Ireland. (Studies in Irish Quaternary deposits, No. 6.) *Proc. R. Ir. Acad., Sect. B* **52B**, 291–314.

Pilgrim, G. E. (1941). The relationship of certain variant fossil types of "horn" to those of the living Pecora. *Ann. Mag. Nat. Hist.* [11] **7**, 172–184.

Rütimeyer, L. (1881). Beiträge zu einer natürlichen Geschichte der Hirsche. *Abh. Schweiz. Paläontol. Ges.* **8**, 3–97.

Schindewolf, O. H. (1946). Zur Kritik des "Biogenetischen Grundgesetzes." *Naturwissenschaften* **8**, 244–249.

Scott, W. B. (1885). Cervalces americanus, a fossil moose, or elk, from the Quaternary of New Jersey. *Proc. Acad. Nat. Sci. Philadelphia* **37**, 181–202.

Scott, W. B., and Osborn, H. F. (1890). Preliminary account of the fossil mammals from the White River and Loup-Fork formations, contained in the Museum of Comparative Zoology. Part II. *Bull. Mus. Comp. Zool.* **20**, 65–100.

Stehlin, H. G. (1937). Bemerkungen über die miocaenen Hirschgenera *Stephanocemas* und *Lagomeryx*. *Verh. Naturforsch. Ges. Basel* **48**, 193–214.

Stehlin, H. G. (1939). *Dicroceros elegans* Lartet und sein Geweihwechsel. *Eclogae Geol. Helv.* **32**, 162–179.

Thenius, E. (1958). Geweihjugendstadien des eiszeitlichen Riesenhirsches, *Megaceros gigan-*

teus (Blum.), und ihre phylogenetische Bedeutung. *Acta Zool. Cracov.* **2,** 707–721.

Voorhies, M. R. (1969). Taphonomy and population dynamics of an early Pliocene vertebrate fauna, Knox County, Nebraska. *Contrib. Geol., Spec. Pap. No. 1,* 1–69.

Webb, S. D. (1973). Pliocene pronghorns of Florida. *J. Mammal.* **54,** 203–221.

Whitworth, T. (1958). "Miocene ruminants of East Africa," Fossil Mammals of Africa, No. 15. Br. Mus. (Nat. Hist.), London.

The Social Significance
of Antlers

The behavior of animals, like that of people, is as species specific as is the form of their bodies. This is particularly true with reference to reproductive activities, which in deer represent highly ritualized behavioral patterns. It is by analyzing the use of antlers by deer in this context that we may begin to appreciate, perhaps even more than the deer themselves, the true functions of these status symbols. Equally important is an understanding of the ecological pressures responsible for the evolutionary arms race that resulted in such luxurious examples of sexual dimorphism.

Rutting Behavior

Overt fighting with antlers, for which deer are famous, is the culmination of a broad repertoire of behavioral patterns that recur every year in response to the seasonal surge in testosterone secretion. As Bubenik (1982) has so aptly perceived, puberty is annually recapitulated by deer in their physiological preparations for rut.

Fig. 64. Hoof flailing by red deer stags after casting their antlers.

Even before the velvet has been shed from the antlers, bucks begin to establish their position in the social hierarchy. This they do by threats with their growing antlers (Dodds, 1958), and occasionally by rearing up on their hind legs and fighting with their front hooves (Fig. 64), a type of aggression also used to good effect by females.

Once the velvet begins to shed, the process is facilitated by rubbing the antlers against solid objects, a practice which in the Roosevelt elk (wapiti) has been interpreted as a means of marking territorial boundaries, presumably in part by leaving their scents on tree trunks (Graf, 1956). Deer have occasionally even been observed to chew off strips of velvet from themselves or each other (Dixon, 1934). However, after most of the velvet has peeled off, various species of deer engage in vigorous threshing activities directed against shrubs and saplings. Species such as wapiti, caribou, and mule deer take out their frustrations on the underbrush, which in the process is mercilessly slashed as it is whipped back and forth by the antlers (Altmann, 1959; Graf, 1956; Pruitt, 1966; Struhsaker, 1967). What purpose this serves can only be surmised, although it has been speculated that this and subsequent sparring activity between males may enable deer to "learn" the shapes and sizes of their antlers (Bubenik, 1968).

Still another prerut activity is wallowing. Typical of larger species of deer, such as moose and wapiti, wallows, or rutting pits, are dug out by the males early in the mating season (Geist, 1963; Thompson, 1949; Woodin, 1956). As these wallows become filled with water, augmented by urine, the deer roll around in them from time to time, plastering their bodies with urine-flavored mud as if to satisfy some inner urge. Even females may occasionally partake of a mud bath in such wallows (Altmann, 1959; Thompson, 1949). Père David's deer have been observed to use their antlers to deposit blobs of mud on their backs (Bedford, 1951).

As a prelude to true fighting, males often engage in sparring contests during the early phases of the rutting season (Barrette, 1977; Bubenik, 1968; Dixon, 1934; Geist, 1974a, 1981). In these encounters, the participants go through the motions of fighting without exhibiting the level of aggression expected in serious combat. Heads are lowered, antler contact is established, but the impression is that the deer are perhaps doing nothing more than measuring their antlers and testing their efficacy (Fig. 65). Such sparring is common in the early fall before the rutting peak, as well as in the spring before the antlers are cast (Franklin and Lieb, 1979). Perhaps it coincides with intermediate levels of testosterone, whether they are on the increase or decrease.

Unlike other ungulates, only muntjacs and roe deer are definitely known to establish territories during the mating season. Roe deer pair off in their territories (Henning, 1962; Kurt, 1968), and may occasionally chase each other in circles that are usually located around trees or bushes. The frequency with which these circles are used is indicated by the depth of the path that may be worn in the ground. Although "roe rings" may figure in courtship behavior, their uses are not restricted to heterosexual pairs or to the mating season alone (Millais, 1897; Schmidt, 1965; "Snaffle," 1904). Muntjacs are also very territorial deer, the males taking considerable care to mark the boundaries with the secretions from their preorbital glands (Dubost, 1971). Père David's deer and fallow deer tend to gather harems, usually in the shade of a large tree (Bedford, 1952; Chaplin and White, 1970). Here the

Fig. 65.   Sparring by mule deer bucks is antecedent to fighting for keeps. (Courtesy of Dr. Valerius Geist.)

"master buck" keeps the less dominant males at a respectful distance while more and more females join the ranks of those already affiliated with such an obviously dominant male. One draws the conclusion that virtually all of the fawns born in such a herd each year may have a single father. Red deer (Bützler, 1974; Clutton-Brock et al., 1979), and to a lesser extent wapiti (Struhsaker, 1967), may also acquire harems, but these tend not to be so restricted to specific localities. Instead, the stag keeps his hinds herded together as best he can while driving off would be suitors. "Boundaries must be arbitrary and subject to considerable alteration, but they are exceptionally sharply defined psychologically at any single moment. These territories are held by the intensity of the stag's sexual jealousy" (Darling, 1937).

Vocalization is often an important releaser of stereotyped social reactions. The bugling of wapiti and roaring of red deer are conspicuous examples of this behavior (Fig. 66). Studies of red deer on the Isle of Rhum, off the Scottish coast, have shown that roaring contests between stags tend to take place when the two individuals are of more or less the same size (Clutton-Brock and Albon, 1979). Apparently the potential antagonism between obviously smaller and larger stags is settled visually before escalating to the roaring stage. Not surprisingly, roaring can be induced by the administration of sex hormones (Fletcher and Short, 1974).

Fig. 66. Red deer stag roaring.

Still another reaction, the significance of which is not fully understood, is the tendency for aggressive males to curl back their upper lips (Geist, 1963). In other animals, sneers are obviously designed to display the upper teeth, particularly the canines. Because deer lack upper incisors, and the canines, when present as tusks, are more laterally placed, it does not seem likely that lip curl, or flehmen, serves this purpose. Observations indicate that flehmen is often, though not always, associated with the male's reaction to the smell of the female, particularly her urine. Olfactory signals are extremely important among deer, particularly with the need for the male to determine by her odor when a female is in estrus. Therefore, flehmen may be a nasal reaction designed to enhance the detection of pheromones.

The onset of fighting in earnest commences with the encounter between two relatively evenly matched males. Such individuals find it useful to estimate the prowess of their opponents before risking a fight. Roaring contests between red deer stags are part of this assessment (Clutton-Brock and Albon, 1979). Visually, the prospective combatants judge each other by walking in parallel at a safe distance (Clutton-Brock and Albon, 1979). The parallel walk has also been described in wapiti by Geist (1966c). This is not unlike the activity observed in mule deer who circle each other before joining combat (Geist, 1981). In either case, individuals expose their maximum dimensions to each other before coming to blows. In reindeer and caribou, the antlers are conspicuously displayed not only by a head-on approach, but also by tilting the head to expose a broadside view of the antlers (Bubenik, 1975, 1982). It has been suggested that the palmate antlers of fallow deer, as well as moose and perhaps the giant deer, *Megaceros,* evolved more for the purpose of intimidating prospective opponents than clashing with them head-on (Gould, 1973, 1974). Lincoln (1972) has suggested the same for red deer.

Compared with the rather protracted preliminary rituals, the fight itself is often a relatively brief encounter. The two deer rush at one another, attempting to effect a blow to the opponent's flank (Geist, 1966b). Each usually catches the other's antlers with his own, whereupon a shoving match ensues. At this stage of the encounter, strength and weight of the body may influence the outcome, enabling the heavier and stronger deer to push his opponent backward. It is the complex branching pattern of antlers that provides the purchase needed to test the strength of the contestants (Geist, 1966b). In red deer and reindeer, the end branches are where the antlers bind together. In deer with less formidable antlers, they may lock like crossed sabers as far down as the brow tines. The latter are well designed not only to inflict injuries on the opponent's head, but also as a protection against such thrusts (Bubenik, 1975). An unfortunate consequence of possessing elaborate antlers is the risk of locking irreversibly with those of an opponent. Unable to disengage their antlers either by pulling or pushing, such deer exhaust themselves in hours or days of futile struggle, only to die of starvation, broken necks, or as victims of opportunistic predators. Such tragic outcomes have been recorded in caribou, moose, Père David's deer, and members of the genus *Odocoileus,* but it undoubtedly happens in other species as well.

Although the role of antlers as display organs cannot be denied, their ultimate purpose is to enable competing males to test each other's dominance in battle. They are obviously not intended as lethal weapons because their architecture does not lend itself to mortal combat the way an anteriorly directed unbranched rapierlike antler would, for example. Nevertheless, there are no holds barred in the combat between aggressive bucks. They go for the throat, chest, or belly, where a punctured lung or herniated intestine will seal the fate of a victim (Geist, 1966b). Among red deer stags, as many as 6% are seriously injured each year (Clutton-Brock et al., 1979). Over 10% of all male mule deer beyond the yearling stage were found to have been injured (Geist, 1974a). Puncture wounds in the skin are the most common lesions inflicted upon an opponent, but broken ribs are not uncommon, and some deer become lame. Such animals are not only eliminated from contention for available females, but run a serious risk of falling victim to predators. Little wonder that the largest animals with the most formidable antlers survive the least number of years. "Display and non-damaging fighting forms," wrote Geist (1966b), "are selected since overt fighters do not live long." While longevity tends to enhance the risk of injury, the evolution of even larger antlers for self-defense is thus favored (Clutton-Brock et al., 1980).

The antlers themselves are in jeopardy of sustaining major damage. It is remarkable how much punishment they can take without breaking (Currey, 1979; Henshaw, 1971), but the frequency with which tines are lost (Bubenik, 1975) testifies to the powerful clashes to which they are subjected. Analysis of antler bone confirms that it is designed to withstand considerable abuse. The outer cylinder of compact bone coupled with the more spongy nature of the core imparts maximal strength. Further, the greater fluid content of antler bone in the autumn endows it with the maximum impact resistance (Chapman, 1981). By spring, when antlers are due to be cast, the water content decreases considerably, rendering them more brittle at a time of year when structural strength is less crucial.

In view of the importance of antlers for display, and the insistence of males to establish their place in the social hierarchy of the herd, it is of some importance to learn to what extent antler size determines social rank. If dominance can be established through display, rather than overt fighting, then energy is conserved and the danger of injury is diminished. Accordingly, Geist (1966a) showed that among bighorn sheep dominance was determined not by body size but by the dimensions of the horns. This is consistent with the observations of Hediger (1946) and Bützler (1974) who found that when deer cast their antlers their rank in the hierarchy declined, but could be regained as new ones grew. Lincoln et al. (1970) noted that the more dominant red deer were those carrying the largest antlers, a correlation also reported by Espmark (1964) for reindeer and by Bergerud (1974) for caribou. If substantial portions of their antlers are cut off, red deer tend to drop in social position (Lincoln, 1972; Lincoln et al., 1970). Studies of castrated red deer by Topiński (1974) showed that when parts of the antlers of castrated stags were lost because of freezing in the winter their relative dominance decreased accordingly, although the process could be reversed as new antlers were grown in the spring and summer.

However, Cameron (1892) wrote that,

> the power of charm, as influencing female choice, finds no place with the Cervidae, because the most vigorous males take forceable possession; and the law of battle which assures the mastery to the strong, favours not the heaviest antlers but the heaviest deer. It is bodily weight that tells, irrespective of the calibre of the weapon.

Observations of reindeer in Alaska, following amputation of their antlers while in velvet, have led to a similar conclusion (Prowse et al., 1980). Premature loss of antlers did not necessarily affect the dominance of these reindeer in either sex. Social status appeared to be determined by size, age, and sexual readiness. Recent studies of red deer stags have even suggested that social rank may affect the sizes to which antlers grow, rather than the other way around (Bartos and Hyanek, 1983). Clearly, the question of how one's position in the social hierarchy is determined in deer, as in humans, is a complex subject which requires further study.

Phylogeny of Function

If antlered mammals had not evolved, the existence of such improbable creatures would never have been imagined, even in the annals of science fiction. The fact that antlers did indeed evolve has challenged zoologists to explain the adaptive significance of these intriguing headpieces (Clutton-Brock, 1982).

A number of theories of antler function have been proposed, foremost of which is that antlers, like horns, may have developed as weapons for combat. This is supported not only by the well-known tendency for deer to lock horns in the mating season, but by the fact that they are male secondary sex characters, and that their possession is accompanied by the aggressive temperaments of male deer during the rutting season. The morphological configurations of antlers also support the combat theory. Except in young animals, or smaller species, antlers are branched outgrowths, a shape conducive not so much to killing as to pushing and wrestling (Geist, 1966b). The branched configuration of deer antlers would appear to be designed to hook on each other in such a way that direct bodily contact is effectively avoided (Bubenik, 1975). Moreover, the fact that the velvet is shed at the end of the growing season when the antlers turn into solid bony headpieces is consistent with the notion that these appendages, first and foremost, have evolved for the aggressive rituals of the mating season. The annual renewal of antlers would therefore appear to be an adaptation in anticipation of the not-infrequent breakages (Bubenik, 1975) that occur to over-zealous combatants.

An equally appealing theory, not necessarily incompatible with the foregoing, is that antlers may have evolved more for display purposes than for head-to-head combat (Coope, 1973; Gould, 1973, 1974; Lincoln, 1972). Field observations confirm that deer do in fact use their racks as social semaphors, often intimidating prospective opponents before coming to blows. The luxurious growth of antlers, particularly in larger species of deer, accentuates the size of the head in ways analogous to similar exaggerations in horn size, mane development, and swollen necks in other ungulates. The seemingly excessive size and branchings of antlers in many types of deer, coupled sometimes with the occurrence of palmation in fallow deer, moose, and reindeer, is more indicative of a visual role than a belligerent one. The palmate configuration of antlers in the Irish elk, and the fact that they tended to grow out to the side rather than upward or forward, is compelling evidence that at least in these giant deer the antlers developed primarily to impress other males visually (Coope, 1973; Gould, 1973, 1974).

On the negative side, it could be argued that if antlers developed strictly for visual display it would not have been necessary to replace them each year. Such headpieces could have remained permanently viable and covered with skin if they were not intended for fighting or were not vulnerable to injuries of combat. It is perhaps significant that a number of fossil ungulates (e.g., *Syndyoceras, Sivatherium*) possessed cephalic appendages, sometimes of impressive shapes and sizes, that were enveloped in skin and not replaced each year (Figs. 46 and 47). One can only assume that these outgrowths must have been display organs. However, it is no coincidence that such creatures were limited to warmer climates.

In 1968, Stonehouse proposed still another hypothesis to explain the presence of antlers in male deer. Noting that males increase their food consumption and metabolism during the summer months to fortify themselves for the autumn rut, he proposed that antlers might function as thermal radiators releasing excess body heat. It must be admitted that the copious blood flow in antlers does in fact dissipate quantities of heat during the growing period. Antlers are almost hot to the touch while in velvet. The branched configuration of antlers increases the surface area, again a possible adaptation to promote heat loss. Presumably the loss of velvet during the winter prevents heat loss during a time of year when its conservation is important.

Although growing deer antlers are admittedly thermal radiators, there are compelling reasons why this was probably not a factor in their evolution (Geist, 1968). If it were, one might expect that tropical species would possess the largest antlers, but they do not. One might also expect tropical deer to remain permanently in velvet if their antlers were thermal radiators, but they shed their velvet and cast their antlers annually as do deer native to the temperate zone.

There are two types of deer in which the possession of antlers for the dissipation of excess body heat would seem to be maladaptive. One is the reindeer (or caribou) which typically begins to grow antlers in the spring when the Arctic climate is still more wintry than summerlike. These animals live in a harsh environment, and to lose valuable heat when snow and ice still prevail does not make sense. Clearly, other factors must outweigh the disadvantages of heat loss from the surfaces of growing antlers in these deer.

Roe deer and Père David's deer are exceptions to the rule that antlers grow during the summer months. For reasons yet to be explained, they have mating seasons in the summer and grow antlers in the winter. If their antlers had evolved as thermal radiators, there should have been powerful selective factors against replacing them at the coldest time of year, particularly in the northern latitudes where roe deer abound.

A final function of antlers has been suggested by Bubenik (1968), who noted the abundance of sebaceous glands in the skin of antlers, and the tendency for deer sometimes to rub their antlers on scent glands elsewhere on the body, especially on the legs (Espmark, 1971b). Hence, antlers could function as "olfactory projectors" in order better to advertise odors of significance to the species. According to Chard (1958), fallow antlers "have clearly been adapted to aid in the dispersal of scent, and the broader the palm the more effective this may be." Although the secretions of sebaceous glands in the velvet do not appear to have a strong scent, at least to the human nose, one cannot rule out the possibility that this odor might be appreciated by other deer. The scent glands on the hind legs of deer produce strong smells (particularly during the mating season) which mark boundaries in territorial species. In nonterritorial deer they could be dispensed into the air from appendages far off the ground. In view of the overwhelming importance of the deer's sense of smell, it is easy to believe that antlers could serve a purpose in the dissipation of pheromones whether produced by their own sebaceous glands off season or picked up from scent glands in rut.

Steinbacher (1957) reported that a red deer had been observed to use his antlers to knock apples out of a tree, and that fallow deer obtained acorns in this way. They also make convenient back-scratchers. Antlers may be many things to many deer, but the problem is to determine which function was primary and which ones were secondary. The olfactory projection theory, although probably based on some truth, must clearly be relegated to the category of a secondary function. Thermal regulation can be discarded as a viable explanation for the origin of antlers, despite the fact that heat loss in such highly vascularized and rapidly growing appendages is inevitable. The social display purpose of antlers to establish rank order and obviate conflicts before they begin constitutes an antler function of mounting significance to behavioral biologists. Although it would be difficult to possess a weapon the very sight of which did not threaten other animals, this important function of antlers would appear to be a derivative of the principal excuse for their existence, namely, to serve as weapons in combat. Therefore, the conclusion is inescapable, as Clutton-Brock (1982) has argued, that antlers must have evolved first and foremost as weapons of aggression, despite the fact that certain secondary functions may have been adopted along the way.

Why the Velvet Is Shed

By definition, antlers are subject to loss and regeneration. From whatever ancestral appendages they may have evolved, they could not be considered true antlers until they acquired the capacity to replace lost parts. If antlers are socially significant structures, their functions must be bound up with the remarkable processes by which they are replaced each year.

The casting and replacement of the antlers themselves do not occur unless the antler first dies. In 1898, Nitsche wrote (in translation) that "it is exceptionally peculiar that a process which is initially purely pathological [i.e., necrosis of antlers] becomes a normal, regularly repeating phenomenon vital to the life of the animal sometime during the phylogenetic development of the Cervidae." We assume from observations of living deer that necrosis is induced by the massive ossification that occurs in the fully grown antler, followed by shedding of the velvet (Fig. 67). Therefore, to seek the phylogenetic explanation of antler casting one must look for reasons why the antler dies and the velvet is shed in the first place.

It has been suggested that with the approach of the mating season, male wapiti are inclined to use their newly developed antlers to mark their territories, however loosely defined such territories might be. This "signpost" behavior may take the form of rubbing the antlers against the bark of trees, threshing the underbrush or digging in the ground to make wallows. In the wapiti, this behavior commonly coincides with the shedding of the velvet, allowing the possible conclusion that the latter is perhaps incidental to the primary activity of staking a claim on a given territory (Graf, 1956). One cannot categorically exclude this as an original reason for shedding the velvet in ancestral forms of deer, but it is important to recognize that not all of the facts are consistent with this notion. For example, loss of velvet is causally related to the ossification of antler bone, a phenomenon not necessarily connected with the marking of territory. Further, most species of deer are not territorial, but they shed their velvet anyway.

A more popular explanation for the loss of velvet relates to the possibility that the precursors of modern antlers may have been vulnerable to injury, presumably inflicted on each other by competing males. One can imagine that this kind of trauma might have favored the selection of individuals capable of healing their wounds most efficiently. Indeed, anyone who has worked with deer cannot help but be impressed with their high degree of

Fig. 67. Fallow deer shedding the velvet, which hangs in tatters from his nearly cleaned antlers.

immunological competence, a capacity so well developed that even abdominal operations can be performed under nonsterile conditions without serious risk to the survival of the animal. Possibly the resistance of deer (and other forms prone to combat) evolved along with the habit of inflicting wounds on each other. Natural selection would have long since acted against those individuals of the population lacking the capacity to overcome infections.

If ancestral deer were predisposed to fight among themselves with viable skin-covered appendages, they must have engaged in some very bloody encounters indeed. However important wound healing must have been to such animals, the avoidance of injuries in the first place would have been sound preventive medicine. Ossification of antlers and spontaneous shedding of the velvet as if in anticipation of the fights to come could have facilitated even more vehement rituals of the rutting season with a minimum of bloodshed. The concomitant demise of the antlers set the stage for their subsequent replacement.

Theories of Antler Casting

One of the earliest theories to explain why deer antlers are lost every year was proposed by the eighteenth century anatomist, John Hunter. "Why do all the deer-tribe cast their antlers?" he wrote. "They do not appear to break off so asily [*sic*] or commonly as to require a yearly renewal. Is it because the buck might be mischievous to the fawn, and he, therefore, is rendered inoffensive at that season?" This astute observation is consistent with the habits and life cycles of most species of deer. They typically cast their antlers in the winter or spring before the fawns are born, and are in velvet during the time of year when the fawns are suckling. Not until after weaning do the adult males shed their velvet and acquire hard, bony antlers.

There are several exceptions to this rule. One is the roe buck which grows antlers during the winter, shedding the velvet in the early spring. Therefore, when the spring fawns are born, the adult males have bony antlers. However, it is worth noting that by the time the males become potentially dangerous to their offspring, they no longer keep company with the females, who isolate themselves in the fawning season (Kurt, 1968). This would be consistent with what Edward, Second Duke of York (1909), noted in "The Master of Game" (believed to have been written originally between 1406 and 1413): "At the bucking of the roebuck he hath to do but with one female for all the season, and a male and a female abide together as the hinds [birds?] till the time that the female shall have her kids; and then the female parteth from the male and goeth to kid her kids far from thence, for the male would slay the young if he could find them." Therefore, because of this separation the newborn fawns are in little danger of being molested by their fathers. By the midsummer mating season the fawns are old enough to fend for themselves.

Père David's deer, like the roe deer, grows antlers throughout the winter, shedding the velvet in the spring at about the time the fawns are born (Table I). How this potential incompatibility is resolved is not known, nor can it be studied in the absence of wild populations to observe.

Another exception is found in tropical deer (Chapter 11). Here the males do not grow their antlers seasonally and in unison, as in temperate zone species. There are males with bony antlers at all times of year, as there are ones in velvet, despite the fact that each individual follows his own 12-month cycle. Because fawns can be born at any time of year, there are always potentially dangerous adult males around. However, there is no evidence that such males pose a threat to fawns. Perhaps behavioral constraints are as effective as developmental ones to ensure the safety of the young animals, as must indeed be the case with the many species of horned ungulates which possess nonrenewable weapons year-round. Nevertheless, the fact that most deer lose their antlers before the fawns are born is probably no coincidence.

Another reason why antlers are lost and replaced each year is to enable their size to keep pace with bodily growth (R. D. Estes, personal communication). Like teeth, antlers mature into nongrowing structures before the animal attains its full stature. Only by being deciduous can they be replaced from time to time with new ones. Horns solve the problem by retaining the ability to elongate, as do continuously erupting incisors and tusks. Because most species of deer require at least several years to mature, it is convenient that their antlers can be replaced by larger and larger racks, thus avoiding allometric embarrassments. In a kind of ontogenetic atavism, they continue to be replaced even after the deer has long since ceased to grow.

The most obvious explanation for replacing antlers is to compensate for anticipated breakage. Although it has been claimed that antler breakage is too rare a phenomenon in nature to account for the need to regenerate (Henshaw, 1971), even a low frequency of annual damage would yield a high cumulative incidence of defective antlers over the lifetime of the individual. Spinage (1970) pointed out that antler replacement might be related to the fact that they are brittle (unlike horns) and therefore subject to frequent breakage.

It is conceivable that repeated injuries to cephalic appendages, particularly if they were secondary sex characters in males, might have led to the evolution of regenerative capacities. Brooke (1878) speculated that

> At first these outgrowths from the frontals remained persistently attached, but eventually the great advantage enjoyed by individuals who through necrosis lost, and through an inherited tendency to produce frontal processes renewed their antlers, over individuals who retained antlers broken and rendered useless by frequent combats, caused the natural selection of the former in the struggle for existence.

Any mutation endowing an individual with powers of regrowth would thus have conferred a selective advantage. Whether or not such regenerative capacities were originally triggered directly by the breakage of the ends of these appendages, or evolved as a spontaneous event recurring each year, cannot be determined from the fossil record. Whatever the answer, it must have been the necrosis of the ossifying antler that set the stage for the development of regenerative capacities in deer, capacities that could not have evolved unless there were something to be replaced.

Another theory concerning the origin of antler casting, first suggested over a century ago (Dawkins, 1878), has been amplified by Coope (1968). This is based on the logical assumption that the primitive antlers of ancestral deer inhabiting temperate zones would have been vulnerable to freezing in the winter. If the ends of their pedicles were seasonally frozen, necrosis would ensue and the dead bone would be separated from the living stump by osteoclastic action in the spring. This is in fact what happens to the antlers of castrated deer (Chapter 13). Because such antlers retain their blood supply, varying lengths of their ends become frozen, depending on the severity of the winter. When spring arrives, the dead portions drop off. Indeed, it has been suggested that the nearly normal detachment of antlers by castrated reindeer, antlers that retain their velvet, could be explained as a consequence of freezing as much as the result of decreased testosterone concentrations.

If such forerunners of modern antlers were lost each year as a result of the temperate habitat, it is possible that in due course of seasonality of this event might well have become linked to the annual recurrence of the reproductive cycle. Indeed, the only way to prevent freezing would have been to stop the blood supply to the antlers before the onset of winter. What could be more logical than to link this event to the rising testosterone levels associated with the autumn breeding season?

It was suggested by Flerov (1952) that deer originated in the tropics. This conclusion was based on the idea that the summer coat of temperate zone deer is the more primitive pelage because it most closely resembles the coloration of newborn fawns. He contends that the winter coat evolved as a secondary adaptation to temperate zone habitats. The similarity between the spotted, often reddish, fur of the fawn and the lighter coloration of the adult summer coat in many species of deer is striking, particularly in the sika deer which tends to be spotted in the warmer months. However, it could be argued that this similarity might be attributed to the fact that fawns possess summer coats only because they are born in the spring or early summer anyway. It is also worth noting that when deer are in their summer coats they are infertile, and the winter coloration is correlated with the fertile phase of the year. Although tropical species of deer may possess lighter, and even spotted, coats (e.g., axis deer), this is obviously not in conflict with their fertility. The earliest fossil Cervidae have been unearthed in Asia and Europe, regions of the world where temperate climates currently prevail. However, it is believed that Miocene climates were more moderate than present ones (Kurtén, 1972). If so, ancestral deer may indeed have evolved at least in neotropical regions where skin-covered cephalic appendages could have been possessed before subsequent changes in climate brought about their conversion to true antlers. Deer may then have invaded the tropics secondarily.

A final explanation for the casting of antlers, if not of the horns of pronghorn antelopes, has been suggested by Geist (1977, 1981) and Geist and Bromley (1978). Noting that the males of many species of deer become thoroughly exhausted and emaciated following an arduous rutting season, and that predators tend to seek out such weakened males for the kill, it was proposed that males may lose their antlers in order to mimic the normally antlerless females. This would make the more vulnerable animals less readily recognized by predators looking for easy game. This ingenious hypothesis is plausible for most species of deer, but it also has some drawbacks, not least of which is the fact that saving males at the expense of pregnant females is not necessarily of benefit to the long-range preservation of the species. Further, one of the most heavily predated kinds of deer is the caribou. However, in this species both sexes carry antlers. The fact that the males cast theirs early in the winter makes them all the more easily distinguishable from the still antlered females. The latter retain their antlers until calving time in the spring. One wonders if the precocious loss of antlers by males and by barren females might decoy predators away from the pregnant cows, a strategy of obvious advantage to the species, males to the contrary notwithstanding.

It would seem equally probable that the early loss of antlers by adult bulls might simply be an economy measure to get rid of cumbersome and energy-consuming structures that have fulfilled their functions. Otherwise, they would have to be carried through the winter when it is more important to conserve energy for survival in a harsh environment.

Perhaps it is more logical to ask not why the male reindeer and caribou drop their antlers so early, but why the females retain them so long. One theory is that this gives the females a competitive edge over other animals in the herd, especially during the most difficult time of year (Espmark, 1971a,b; Henshaw, 1968, 1969).Elevation of the social status of female reindeer and caribou could have two effects. One would be to give last year's calves an advantage in the social hierarchy, although the early postnatal growth of antlers by young animals of this species and their relative independence toward the end of their first year of life might make it unnecessary for them to benefit from the protection of their mothers. A more logical explanation, therefore, is that reindeer and caribou cows require all the advantages they can get in order to protect themselves from predators and compete effectively for limited nutrients at a time of year when their unborn fetuses are undergoing such rapid growth. The fact that barren females lose their antlers significantly earlier in the season than do pregnant ones supports this hypothesis.

Ultimate Explanations

The morphology of antlers, like that of other display organs, must be coordinated with appropriate patterns of behavior if their ultimate function is to be served (Geist, 1966b). Therefore, structure and function have evolved together, each having been shaped by ecological interactions favoring the perpetuation of those individuals in greatest harmony with their surroundings, both environmental and social. Consideration of the social functions of antlers can be enhanced by a broader appreciation of display organs in general and their relations to reproductive behavior in a wider spectrum of animals. Antlers are a special case of the general phenomenon whereby various creatures advertise themselves and attract or intimidate their conspecifics.

Although the mere presence of an animal is sometimes communication enough, its appearance and behavior comprise a body language that is only now being translated by ethologists and ecologists. Animal bodies are adorned with meaning. Their coloration may be sexually dimorphic or monomorphic, cryptic or conspicuous. Their sizes, whether large or small, are related to their diets, habitats, and degrees of gregariousness. The antlers of deer and the horns and tusks of other ungulates serve to accentuate the head, and are sometimes enhanced by manes, beards, or dewlaps (Geist, 1966b). Notwithstanding the attention-getting attributes of the tails and bright rump patches of some deer and antelopes, there has been a definite trend toward anterior differentiation in ungulates, a trend that has been paralleled by the evolution of complimentary behavior patterns.

In view of the sexual dimorphism that antlers represent in all but reindeer and caribou, it is interesting to note that sexually dimorphic ungulates in general are usually small, solitary browsers that prefer concealed habitats and are cryptically colored (Alexander *et al.*, 1979; Estes, 1974). They not only hide from danger themselves, but also conceal their newborn young. If resources are plentiful, monogamous pair bonds may form the basis for territorial behavior. Some of the smaller species of ungulates are included in this category, animals that carry relatively short horns or antlers or possess tusklike canines in the males. In contrast, sexually monomorphic ungulates are often large, gregarious, migratory grazers that prefer open habitats where they can run from danger, and expect their precocial young to do likewise. They may have conspicuous coloration and large, well-developed horns or antlers in both sexes. If the male tends to be larger than the female, it is the result of the protracted period of maturation. Such males may require a number of years to reach sexual maturity, during which time they continue to grow larger each year and to develop impressive horns or antlers. Adolescent males tend to resemble adult females, and their relatively delayed sexual maturity has appropriately been referred to as an example of neoteny (Estes, 1974). In more socially advanced ungulates, the males may establish dominance hierarchies, sometimes as the basis for territorial behavior during the mating season. Gregarious ungulates, like schools of fishes and flocks of birds, find their safety in being lost in a crowd. Solitary ones seek the safety of dense cover. In either case, the advantages of interposing objects, be they other individuals or vegetation, between themselves and predators seem not to be unappreciated by most herbivores.

Deer fit into the overall scheme of ungulate behavioral ecology in a number of ways. The deer that are most gregarious in their habitats, i.e., reindeer and caribou, are migratory animals that do not hide their young and possess antlers in both sexes. It is interesting to note that the percentage of females lacking antlers is greater in woodland than in barren ground caribou. However, like sexually dimorphic deer they tend to be polygynous. In fact, monogamy is virtually unknown among deer, with the exception of the annual pair bonds that may be established by roe deer.

Finally, the question of what has been responsible for the evolution of such magnificent antlers as have evolved in temperate climates remains to be answered. It seems unlikely that they may have been selected in response to predators (Geist, 1966b), if only because antlers are clearly used as offensive weapons against other conspecific males, and only as a last resort as defenses against animals of prey.

It is probably no coincidence that the largest antlers have been grown by deer inhabiting previously glaciated lands. According to Geist's (1974b) dispersal theory, deer invaded such habitats where plentiful resources relatively free from competition permitted maximal growth of individuals and favored the establishment of territoriality. This led to the evolution of larger deer, which in turn required bigger antlers with which to defend their territories. Allometric dictates enhanced progressively greater cephalic polarization, culminating in *Megaceros* and other giant deer. Similarly, the huge racks of antlers grown by reindeer, caribou, moose, and the various kinds of red deer and wapiti testify to the operation of powerful ecological influences during the late Pliocene and Pleistocene epochs.

Coupled with the opportunities created by receding glaciers were the effects of the temperate climate on cervid evolution (Geist, 1974b). Whereas tropical species are unaffected by alternating summers and winters, deer living in higher latitudes must not only synchronize their reproduction with the seasons, but ensure that their fawns are born not too early to become the victims of late spring snowstorms, nor so late that they cannot fend for themselves by early autumn. Correspondingly, the rutting season becomes increasingly shortened at progressively higher latitudes (Chapter 11), intensifying the aggressive behavior of males in competition for females whose estrous cycles are focused in a relatively brief period of time. While the frenzy of the rutting period may be accentuated, it is at least not spread out over the entire year as is the case with tropical deer. Such a pattern has the advantage of allowing the males a period of grace from their annual reproductive obligations, enabling them to spend the spring and summer months growing antlers, depositing fat, and increasing their body mass, unencumbered by the demands of producing sperm and reacting to the tyranny of testosterone.

References

Alexander, R. D., Hoogland, J. L., Howard, R. D., Noonan, K. M., and Sherman, P. W. (1979). Sexual dimorphisms and breeding systems in pinnipeds, ungulates, primates, and humans. In "Evolutionary Biology and Human Social Behaviour" (N. Chagnon and W. Irons, eds.), pp. 402–435. Duxbury Press, North Scituate, Massachusetts.
Altmann, M. (1959). Group dynamics in Wyoming moose during the rutting season. J. Mammal. 40, 420–424.
Barrette, C. (1977). Fighting behaviour of muntjac and the evolution of antlers. Evolution 31, 169–176.
Bartos, L., and Hyanek, J. (1983). The influence of social position on antler growth in the red deer stag. In "Antler Development in Cervidae" (R. D. Brown, ed.), Caesar Kelberg Wildl. Res. Inst., Texas A & I University, Kingsville (in press).
Bedford, Duke of (1951). Père David's deer: The history of the Woburn herd. Proc. Zool. Soc. London 121, Part II, 327–333.
Bedford, Duke of (1952). Père David's deer. Zoo Life 7, 47–49.
Bergerud, A. T. (1974). Rutting behaviour of Newfoundland caribou. In "The Behaviour of Ungulates and its Relation to Management" (V. Geist and E. Walther, eds.), New Ser. Publ. 24, Vol. I, pp. 395–435. IUCN, Morges, Switzerland.
Brooke, V. (1878). On the classification of the Cervidae, with a synopsis of the existing species. Proc. Zool. Soc. London pp. 883–928.
Bubenik, A. B. (1968). The significance of the antlers in the social life of the Cervidae. Deer 1, 208–214.
Bubenik, A. B. (1975). Significance of antlers in the social life of barren ground caribou. Proc. Int. Reindeer Caribou Symp., 1st, 1972 pp. 436–461.
Bubenik, A. B. (1983). The behavioral aspects of antlerogenesis. In "Antler Development in Cervidae" (R. D. Brown, ed.), Caesar Kleberg Wildl. Res. Inst., Texas A & I University, Kingsville (in press).
Bützler, W. (1974). Kampf- und Paarungsverhalten, soziale Rangordnung und Aktivitätsperiodik beim Rothirsch (Cervus elaphus L.). J. Comp. Ethol., Suppl. 16, 1–80.
Cameron, A. G. (1892). The value of the antlers in the classification of deer. Field 79, 625.
Chaplin, R. E., and White, R. W. (1970). The sexual cycle and associated behaviour patterns in the fallow deer. Deer 2, 561–565.
Chapman, D. I. (1981). Antler structure and function—a hypothesis. J. Biomech. 14, 195–197.
Chard, J. S. R. (1958). The significance of antlers. Bull. Mammal. Soc. Br. Isles 10, 9–14.
Clutton-Brock, T. H. (1982). The functions of antlers. Behaviour 79, 108–125.
Clutton-Brock, T. H., and Albon, S. D. (1979). The roaring of red deer and the evolution of honest advertisement. Behaviour 69, 145–170.
Clutton-Brock, T. H., Albon, S. D., Gibson, R. M., and Guinness, F. E. (1979). The logical stag: Adaptive aspects of fighting in red deer (Cervus elaphus L.). Anim. Behav. 27, 211–225.

Clutton-Brock, T. H., Albon, S. D., and Harvey, P. H. (1980). Antlers, body size and breeding group size in the Cervidae. *Nature (London)* **285**, 565–567.

Coope, G. R. (1968). The evolutionary origins of antlers. *Deer* **1**, 215–217.

Coope, G. R. (1973). The ancient world of "Megaceros." *Deer* **2**, 974–977.

Currey, J. D. (1979). Mechanical properties of bone tissues with greatly differing functions. *J. Biomech.* **12**, 313–319.

Darling, F. F. (1937). "A Herd of Red Deer. A Study of Animal Behaviour." Oxford Univ. Press, London/New York.

Dawkins, W. B. (1878). Contributions to the history of the deer of the European Miocene and Pliocene strata. *Q. J. Geol. Sci. London* **34**, 402–420.

Dixon, J. S. (1934). A study of the life history and food habits of mule deer in California. Part I. Life history. *Calif. Fish Game* **20**(3), 181–282.

Dodds, D. G. (1958). Observations of pre-rutting behavior in Newfoundland moose. *J. Mammal.* **39**, 412–416.

Dubost, G. (1971). Observations éthologiques sur le Muntjak (*Muntiacus muntjak* Zimmerman 1780 et *M. reevesi* Ogilby 1839) en captivité et semi-liberté. *Z. Tierpsychol.* **28**, 387–427.

Edward, Second Duke of York (1909). *In* "The Master of Game" (W. A. Baillie-Grohman and F. Baillie-Grohman, eds.), p. 41. Chatto & Windus, London.

Espmark, Y. (1964). Studies in dominance-subordination relationship in a group of semi-domestic reindeer (*Rangifer tarandus* L.). *Anim. Behav.* **12**, 420–426.

Espmark, Y. (1971a). Antler shedding in relation to parturition in female reindeer. *J. Wildl. Manage.* **35**, 175–177.

Espmark, Y. (1971b). Mother-young relationship and ontogeny of behaviour in reindeer (*Rangifer tarandus* L.). *Z. Tierpsychol.* **29**, 42–81.

Estes, R. D. (1974). Social organization of the African Bovidae. *In* "The Behaviour of Ungulates and Its Relation to Management" (V. Geist and F. Walther, eds.), New Ser. Publ. 24, Vol. I, pp. 166–205. IUCN, Morges, Switzerland.

Flerov, K. K. (1952). Musk deer and deer. *In* "Fauna of USSR. Mammals," Acad. Sci. USSR, Moscow.

Fletcher, T. J., and Short, R. V. (1974). Restoration of libido in castrated red deer stag (*Cervus elaphus*) with oestradiol-17β. *Nature (London)* **248**, 616–618.

Franklin, W. L., and Lieb, J. W. (1979). The social organization of a sedentary population of North American elk: A model for understanding other populations. *In* "North American Elk: Ecology, Behavior and Management" (M. S. Boyce and L. D. Hayden-Wing, eds.), pp. 185–198. University of Wyoming, Laramie.

Geist, V. (1963). On the behaviour of the North American moose (*Alces alces andersoni* Peterson 1950) in British Columbia. *Behaviour* **20**, 377–416.

Geist, V. (1966a). The evolutionary significance of mountain sheep horns. *Evolution* **20**, 558–566.

Geist, V. (1966b). The evolution of horn-like organs. *Behaviour* **27**, 175–214.

Geist, V. (1966c). Ethological observations on some North American cervids. *Zool. Beitr.* [N. S.] **12**, 219–250.

Geist, V. (1968). Horn-like structures as rank symbols, guards and weapons. *Nature (London)* **220**, 813–814.

Geist, V. (1974a). On fighting strategies in animal combat. *Nature (London)* **250**, 354–355.

Geist, V. (1974b). On the relationship of ecology and behaviour in the evolution of ungulates: Theoretical considerations. *In* "The Behaviour of Ungulates and its Relation to Management" (V. Geist and F. Walther, eds.), New Ser. Publ. 24, Vol. I, pp. 235–246. IUCN, Morges, Switzerland.

Geist, V. (1977). A comparison of social adaptations in relation to ecology in gallinaceous bird and ungulate societies. *Annu. Rev. Ecol. Syst.* **8**, 193–207.

Geist, V. (1981). Behavior: Adaptive strategies in mule deer. In "Mule and Black-tailed Deer of North America" (D. C. Wallmo, ed.), pp. 157–223. Univ. of Nebraska Press, Lincoln.

Geist, V., and Bromley, P. T. (1978). Why deer shed antlers. *Z. Saugetierkd.* **43**, 223–231.

Gould, S. J. (1973). Positive allometry of antlers. *Nature (London)* **244**, 375–376.

Gould, S. J. (1974). The origin and function of "bizarre" structures: Antler size and skull size in the "Irish elk," *Megaloceros giganteus. Evolution* **28**, 191–220.

Graf, W. (1956). Territorialism in deer. *J. Mammal.* **37**, 165–170.

Hediger, H. (1946). Zur psychologischen Bedeutung des Hirschgeweihs. *Verh. Schweiz. Naturforsch. Ges.* **126**, 162–163.

Henning, R. (1962). Uber das Revierverhalten der Rehbocke. *Z. Jagdwiss.* **8**, 61–81.

Henshaw, J. (1968). A theory for the occurrence of antlers in females of the genus *Rangifer. Deer* **1**, 222–226.

Henshaw, J. (1969). Antlers—The bones of contention. *Nature (London)* **224**, 1036–1037.

Henshaw, J. (1971). Antlers—The unbrittle bones of contention. *Nature (London)* **231**, 469.

Hunter, J. (1861). In "Essays and Observations on Natural History, Anatomy, Physiology, Psychology and Geology" (R. Owen, ed.), Vol. 2, pp. 136–139. John Van Voorst, London.

Kurt, F. (1968). "Das Sozialverhalten des Rehes *Capreolus capreolus* L. Eine Feldstudie" Parey, Berlin.

Kurtén, B. (1972). "The Age of Mammals." Columbia Univ. Press, London/New York.

Lincoln, G. A. (1972). The role of antlers in the behaviour of red deer. *J. Exp. Zool.* **182**, 233–250.

Lincoln, G. A., Youngson, R. W., and Short, R. V. (1970). The social and sexual behaviour of the red deer stag. *J. Reprod. Fertil., Suppl.* **11**, 71–103.

Millais, J. G. (1897). "British Deer and Their Horns." Henry Sotheran & Co., London.

Nitsche, H. (1898). "Studien über Hirsche." Engelmann, Leipzig.

Prowse, D. L., Trilling, J. S., and Luick, J. R. (1980). Effects of antler removal on mating behavior of reindeer. In "Proceedings of the Second International Reindeer/Caribou Symposium" (E. Reimers, E. Gaare, and S. Skjenneberg, eds.), pp. 528–536. Direktoratet for vilt og ferskvannsfisk, Trondheim.

Pruitt, W. O., Jr. (1966). The function of the brow tine in caribou antlers. *Arctic* **19**, 111–113.

Schmidt, P. (1965). "Das Reh. Sein Leben—Sein Verlalten." Hallwag Verlag, Bern.

"Snaffle," D. R. (1904). "The Roedeer." E. M. Harwar, London.

Spinage, C. A. (1970). Giraffid horns. *Nature (London)* **227**, 735–736.

Steinbacher, G. (1957). Zum Geweihgebrauch der Hirsche. *Saugetierkd. Mitt.* **5**, 75.

Stonehouse, B. (1968). Growing antlers may have a thermoregulatory function. *Nature (London)* **218**, 870–871.

Struhsaker, T. T. (1967). Behavior of elk (*Cervus canadensis*) during the rut. *Z. Tierpsychol.* **24**, 80–114.

Thompson, W. K. (1949). Observations on moose courting behavior. *J. Wildl. Manage.* **13**, 313.

Topiński, P. (1974). The role of antlers in establishment of the red deer herd hierarchy. *Acta Theriol.* **19**, 509–514.

Woodin, H. E. (1956). The appearance of a moose rutting ground. *J. Mammal.* **37**, 458–459.

A Fawn's First Antlers

For obvious reasons, antlers cannot begin to grow until after birth. Yet even in its prenatal development the fetal fawn shows unmistakable signs of things to come. Each frontal bone on the skull grows a barely palpable bump at the sites where antlers are later to appear (Lincoln, 1973). Even females possess them because, although they do not normally grow antlers, they nevertheless are endowed with the potential to do so (Chapter 14). Externally, the locations of future antlers is disclosed by precocious hair development in the fetus (Wika, 1980), and by a pair of cowlicks in the postnatal pelage.

Pedicle Development

Before a deer can produce antlers, it must first grow pedicles, or stumps on the ends of which the antlers can develop (Fig. 68). These pedicles may be several centimeters long in young deer, but they become progressively shorter as the deer matures and they are remodeled into the expanding skull. In adults the antlers appear to be attached directly to the skull. Nevertheless, it is in the pedicle, or in the surrounding parts of the skull or scalp, that the

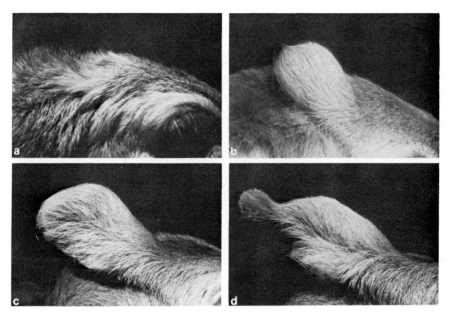

Fig. 68. Stages in the development of a yearling fallow deer's first pedicles and antlers. Photographs were taken on (a) March 16, (b) May 15, (c) July 7, and (d) August 1. The conversion of pedicle tip to antler bud is a gradual process.

tissues from which antlers are regenerated each year are located. It is axiomatic that in order for a missing appendage to be replaced something must be left behind from which the new parts can develop. The pedicle fulfills this role in antler regeneration. In fact, more is known about the origin of the pedicle than of the antler itself (Gruber, 1952).

Histologically, the place where the future pedicle is destined to develop is recognizable only as a thickened region of the fawn's frontal bone. The periosteum overlying this location, which may become 1 mm thick, is noticeably tougher than elsewhere on the cranium, an attribute that facilitates peeling it off intact in surgical transplantation experiments. Further, as Hartwig (1967) first noted in roe deer fawns, a conspicuous fibrous cord, the significance of which remains to be explored, extends in the subcutaneous connective tissue from the skull to the overlying skin in this region.

The incipient antler pedicle differs from the surrounding cranium in that it is composed of spongy bone instead of compact plates. Its trabeculae are augmented apically by the deposition of new ossified tissue derived from the cap of antlerogenic periosterum. In this manner, the pedicle gradually pushes upward beneath the scalp, eventually becoming externally visible instead of just palpable. By this stage of development, chondrogenesis supervenes in the growing apex of the pedicle. Irregular columns of cartilage are formed as extensions of the bony trabeculae, the two types of tissue merging almost imperceptibly into each other. However, occasional multinucleate cells in the region of transition are presumed to be chondroclasts that mediate the replacement of cartilage by bone. The entire structure is richly vascularized.

The manner in which the antler pedicle is formed sets it apart from the mechanisms of horn development (Chapter 3). The latter typically originate in an *os cornu* in the Bovidae, or an ossicone in giraffes. These represent separate centers of bone formation in the subcutaneous connective tissue. Such ossicles secondarily fuse with the bone of the underlying skull, eventually to develop into the bony core of the horn. In the bovids, this horn bone is lightened by air spaces confluent with the sinuses in the skull. In giraffes, the ossicone remains as a relatively solid structure. Antlers are also solid bony appendages when fully mature, but they are derived not as an epiphysis, or isolated bone, but as an apophysis, which is a direct outgrowth from the cranium.

Antlerogenic Transformation

At first, the pedicles are enveloped in the skill and fur of the scalp. They usually remain so until lengths of up to several centimeters are attained. In due course, however, their tips acquire a texture different from that of the scalp itself, a texture more characteristic of the velvety skin in which antlers are normally enveloped. This skin is shiny and more sparsely populated with hairs. The hairs are not the type that lie nearly flat on the surface of the skin as in the case of body fur, but grow straight out at right angles. Each hair follicle is associated with a sebaceous gland from which an oily secretion (sebum) is produced. This sebum is responsible for the shiny appearance of the velvet skin. That it may contain pheromones, or serve as an insect repellent, has been suggested but not proven.

The mechanisms by which the tip of the initial pedicle is transformed into an antler bud have not been thoroughly investigated. Because it is such a gradual process, it is impossible to pinpoint the exact point when pedicle growth gives way to antler development (Gruber, 1952). The genesis of the first antlers is a phenomenon that is not conveniently classified. Although the histological events undoubtedly resemble those by which subsequent sets of antlers are regenerated each year, the process is not an example of regeneration because there has been nothing lost to be replaced. Further, it is not analogous to embryonic limb or tail bud formation. These involve the primary differentiation of embryonic ectoderm and mesenchyme, but the first antler is created out of previously differentiated tissues in the postnatal animal. In this respect, the fawn's initial antler is a unique zoological structure the development of which deserves at least to be described in detail if it is to be understood at all.

Fawns versus Yearlings

Once the antler bud is established, elongation proceeds without delay. In yearlings, the resulting antler is typically an unbranched spike, the length of which varies with the precocity of the deer and the size of the species. They may be mere knobs less than 1 cm high on the tips of the pedicles in deer that were born late in the season or were not optimally nourished. At the other extreme, the production of branched antlers by yearlings has on rare occasions been reported in moose (Cringan, 1955; Peterson, 1955), red deer (Millais, 1906), and wapiti (Murie, 1951). In some white-tailed deer populations, 2-point antlers may be relatively common in yearlings (Harmel, 1983).

Although the majority of deer begin to grow their first sets of antlers in the spring and shed the velvet in September of their second year, there are several species that for unaccountable reasons may produce their first antlers as fawns rather than as yearlings. Reindeer and caribou are the most conspicuous examples of this phenomenon. Calves are born in May as with most other deer, but within a few weeks both sexes begin to produce frontal outgrowths that develop into short spike antlers by the end of their first summer (Cameron, 1892; Harper, 1955; Seton, 1929). Moose also produce antlers as calves, usually little more than "buttons" hidden in the fur (Peterson, 1955), but outgrowths over 10 cm long have been reported in one case (Sugden, 1964). These "infant antlers" grow out during the fall and shed their velvet by late winter (Fig. 69). They drop off as renewed growth commences in April. Diminutive antlers may also be produced by roe bucks during their first autumn (Lebedinsky, 1939; Tegner, 1961; Zimmer, 1905). Similarly, infant antlers are sometimes grown by mule and white-tailed deer in the fall (Anderson and Medin, 1971; Lindsdale and Tomich, 1953). They seldom reach more than 1–2 cm in length, becoming bony buttons in the winter (Dixon, 1934; French et al., 1955).

The reason for this anomalous production of antlers by the fawns of a few species of deer is a matter for conjecture. It may be significant that all of the aforementioned deer are native to high latitudes. One could speculate that the growth of antlers by fawns might be correlated with their more rapid maturation, and that this in turn could be an adaptation to the shorter

Fig. 69. "Button" antler of a yearling moose. The bony remnant (arrow) still remains attached on April 30 while renewed growth begins to mushroom around it. (Courtesy of Eldon Pace.)

summer seasons in the Arctic and northern temperate zone. Closer to the equator there is plenty of room for flexibility in the mating and fawning seasons, with no rush for the offspring to be able to fend for themselves. Farther north there is little latitude for deviation from a seasonal reproductive schedule that must be rigidly adhered to if fawns are to be prepared for a possibly early winter. Thus, such young deer may well have been at an advantage if they were to accelerate their development.

Photoperiod and Hormones

The timing of pedicle and antler production in the first year or so has been shown not to be affected by light cycles, but to be innately programmed into the developmental schedule by which a fawn matures (Goss, 1969, 1980; Sempéré and Lacroix, 1983; Suttie, 1983). Not until their second year do deer become susceptible to the seasonal fluctuations in day length (Chapter 11) that thenceforth regulate the annual cycles of reproductive activity, including the replacement of antlers every year.

It was known even to Aristotle that if a fawn were castrated he would not grow antlers (Chapter 13). Thus, in the absence of testosterone, pedicles are unable to develop from the barely palpable frontal protuberances that are present at birth. It might be concluded, therefore, that the failure of pedicles to grow in females could be attributed to the absence of testosterone. If male sex hormones are injected into female red deer, small pedicles may be induced to grow, but not antlers (Jaczewski, 1981). However, it could also be the presence of the female hormone that normally holds in abeyance a doe's latent potential for pedicle and antler growth, even though their development is not induced by spaying. When estrogen was administered to male sika fawns throughout the period when they would normally be growing their first sets of pedicles and antlers, the production of such structures was completely inhibited (R. J. Goss, unpublished data). Only when treatment was eventually discontinued did these "feminized" males belatedly grow pedicles and antlers. Therefore, the onset of pedicle and antler growth during the first year is under endocrine control, but the secretion of hormones would appear not to occur in response to photoperiodic cycles until the deer's second year.

Histogenesis

If antlers only grow on pedicles, from what tissues do the pedicles develop? Clearly, their production is associated with the frontal bumps and the overlying whorls of hair in the scalp of newborn fawns. However, these structures include a variety of separate tissues such as bone, periosteum, connective tissue, dermis and epidermis, not to mention nerves and blood vessels. One approach designed to narrow the choices is to delete one or another of these histologic candidates. Early experiments (Goss et al., 1964) were conducted on weanling white-tailed fawns. Three different operations were performed on anesthetized animals. In one, a round area of the scalp about 3 cm in diameter was resected from the presumptive antler region on one side. In another group, a flap of scalp was reflected to expose the underlying frontal protuberance. The latter, about 1 cm in diameter, was then cut off even with the surrounding skull, and the flap of skin was then replaced and sutured. In a third procedure, both skin and bone were excised

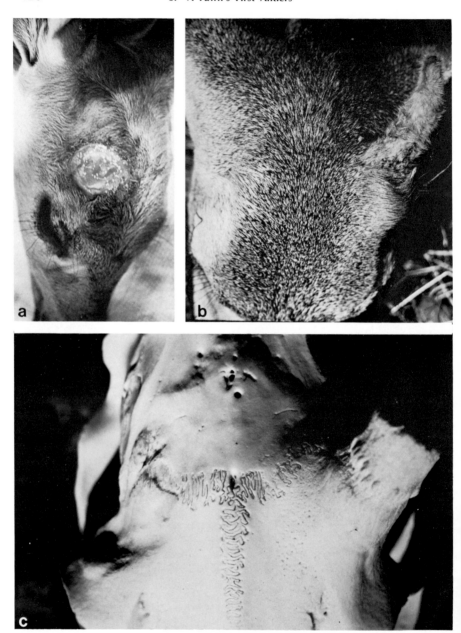

Fig. 70.   (a) White-tailed fawn from which the skin and presumptive antler bone have been excised during the animal's first summer. (b) Same deer after his second summer showing absence of antler on operated (right) side. (c) Skull of 3-year-old deer illustrating absence of pedicle on right side. (From Goss et al., 1964.)

(Fig. 70). After their wounds had healed, the deer were released in a game preserve for later recapture. When these deer were recovered in their second year, their unoperated control sides had grown normal spike antlers. However, on the operated sides, antlers had been produced by those deer from which the skin alone had been removed, but not by those whose bony frontal protuberances had been cut off, with or without the overlying skin (Fig. 70). These findings indicated that it was the bone, not the skin, that must be responsible for the initial development of pedicles, and of the antlers that grow from them.

A more thorough and conclusive series of experiments was later conducted by Hartwig in Germany. This investigator used transplantation techniques to identify the tissue(s) from which pedicles develop. First, parts of the scalp from the prospective antler sites of 3-week-old roe deer fawns were grafted to the thigh. Although the skin survived, no antlers were produced (Hartwig, 1967). In a reciprocal operation, leg skin was transplanted to the denuded area of the forehead, but the graft failed to take. Nevertheless, healing eventually occurred from the surrounding scalp, and there later developed a pedicle and antler that were only slightly shorter than on the opposite unoperated side (Hartwig, 1967). Next, the periosteum was separated from the presumptive antler region of the frontal bone and shifted as a flap graft to a site beneath the skin in the center of the forehead (Hartwig,

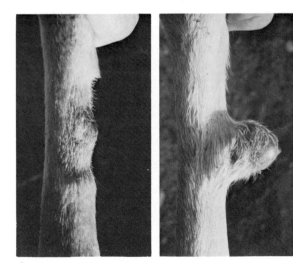

Fig. 71. Induction of ectopic antler from graft of antlerogenic periosteum subcutaneously to the foreleg of a fallow deer fawn. Following operation on January 8, successive stages in development are shown (left to right) on February 14, June 17, and October 15, when the velvet had been shed. Casting and regrowth occurred in subsequent years.

1968). At this location a pedicle and antler subsequently developed, while no such outgrowth formed at the original place deprived of its periosteum. Similar induction of antler growth occurred from the surface of the parietal bone when pedicle periosteum was grafted beneath the skin to this site (Hartwig and Schrudde, 1974). Finally, it was shown that even leg skin can participate in antler growth when pedicle periosteum is grafted between the metacarpal bone of the foreleg and its overlying skin. Antlers induced at this ectopic location were abnormally short, but underwent ossification, shed their velvet, and were replaced in synchrony with the remaining antler on the head (Hartwig and Schrudde, 1974). This remarkable breakthrough not only demonstrates the histogenesis of antler pedicles, but proves that the skin plays a relatively passive role in the initiation of pedicle and antler development in fawns and yearlings.

These results have since been confirmed in experiments on fallow deer by R. J. Goss (unpublished data). Discs of antlerogenic periosteum 1.5 cm in

Fig. 72. A hairless *nude* mouse to which antlerogenic periosteum from a fallow deer fawn had been transplanted beneath the scalp. An ossicle has differentiated at the graft site, but no antler. (Courtesy of Dr. Richmond Prehn.)

diameter will induce pedicle bone and limited antler formation (Fig. 71) when autotransplanted subcutaneously to the leg, hip, or ear of a fawn. Grafts of one-half this diameter do not promote antler development, but may give rise to ossicles, as do intramuscular transplants. Even when implanted beneath the skin of *nude* mice (a hairless and athymic mutant incapable of immunologic rejection), relatively large masses of bone develop, but the overlying skin does not transform into velvet (Fig. 72). It would seem that mice lack the genes for antler development that deer possess.

The foregoing results prove conclusively that pedicles and antlers originate from the frontal periosteum. This represents yet another distinction between horns and antlers, in addition to the different ways in which their respective bony components develop. Antlerogenesis is initiated in the periosteum, and although skin is necessary for the full differentiation of antlers, it only becomes secondarily induced by subjacent skeletogenic events. However, in the case of horns, excision of the skin from the presumptive horn site precludes further development, including that of the *os cornu* (Brandt, 1928; Dove, 1935; Kômura, 1926). Therefore, horns take their origins in the integument proper, which in turn induces appropriate kinds of bone formation in underlying tissues. Antlers derive from the periosteum which transforms the overlying scalp into velvet. Horns and antlers may be superficially similar in appearance and function, but that is where the resemblance ends.

References

Anderson, A. E., and Medin, D. E. (1971). Antler phenology in a Colorado mule deer population. *Southwest. Nat.* **15,** 485–494.

Aristotle. "Historia Animalum" (Eng. transl. by D'A. W. Thompson), p. 631. Oxford Univ. Press (Claredon), London/New York.

Brandt, K. (1928). Die Entwicklung des Hornes beim Rinde bis zum Beginn der Pneumatisation des Hornzapfens. *Morphol. Jahrb.* **60,** 428–468.

Cameron, A. G. (1892). Value of the antlers in classification of deer. *Field* **2055,** 703.

Cringan, A. T. (1955). Studies of moose antler development in relation to age. In "North American Moose" (R. L. Peterson, ed.), pp. 239–247. Univ. of Toronto Press, Toronto.

Dixon, J. S. (1934). A study of the life history and food habits of mule deer in California. Part I. Life history. *Calif. Fish Game* **20**(3), 181–282.

Dove, W. F. (1935). The physiology of horn growth: A study of the morphogenesis, the interaction of tissues, and the evolutionary processes of a Mendelian recessive character by means of transplantation of tissues. *J. Exp. Zool.* **69,** 347–406.

French, C. E., McEwen, L. C., Magruder, N. D., Ingram, R. H., and Swift, R. W. (1955). Nutritional requirements of white-tailed deer for growth and antler development. *Bull.— Pa., Agric. Exp. Stn.* **600,** 1–50.

Goss, R. J. (1969). Photoperiodic control of antler cycles in deer. II. Alterations in amplitude. *J. Exp. Zool.* **171,** 223–234.

Goss, R. J. (1980). Photoperiodic control of antler cycles in deer. V. Reversed seasons. *J. Exp. Zool.* **211,** 101–105.

Goss, R. J., Severinghaus, C. W., and Free, S. (1964). Tissue relationships in the development of pedicles and antlers in the Virginia deer. *J. Mammal.* **45,** 61–68.

Gruber, G. B. (1952). Über das Wesen der Cerviden-Geweihe. *Dtsch. Tierärztl. Wochenschr.* **59,** 225–228, 241–243.

Harmel, D. (1983). The effects of genetics on antler quality in white-tailed deer (Odocoileus virginianus). *In* "Antler Development in Cervidae" (R. D. Brown, ed.), Caesar Kleberg Wildl. Res. Inst., Texas A & I University, Kingsville (in press).

Harper, F. (1955). "The Barren Ground Caribou of Keewatin," University of Kansas, Mus. Nat. Hist., Misc. Publ. No. 6. Allen Press, Lawrence.

Hartwig, H. (1967). Experimentelle Untersuchungen zur Entwicklungsphysiologie der Stangenbildung beim Reh (Capreolus c. capreolus L. 1758). *Wilhelm Roux Arch. Entwicklungs Mech. Org.* **158,** 358–384.

Hartwig, H. (1968). Verhinderung der Rosenstock- und Stangenbildung beim Reh, *Capreolus capreolus,* durch Periostauschaltung. *Zool. Garten (Leipzig)* [N. S.] **35,** 252–255.

Hartwig, H., and Schrudde, J. (1974). Experimentelle Untersuchungen zur Bildung der primären Stirnauswüchse beim Reh (*Capreolus capreolus* L.). *Z. Jagdwiss.* **20,** 1–13.

Jaczewski, Z. (1981). Further observations on the induction of antler growth in red deer females. *Folia Biol. (Krakow)* **29,** 131–140.

Kômura, T. (1926). Transplantation der Hoerner bei Cavicornien und Enthornung nach einfachster Methode. *J. Jpn. Soc. Vet. Sci.* **5,** 69–85.

Lebedinsky, N. G. (1939). Beschleunigung der Geweihmetamorphose beim Reh (Capreolus capreolus, L.) durch das Schilddrüsenhormon. *Acta Biol. Latv.* **9,** 125–132.

Lincoln, G. A. (1973). Appearance of antler pedicles in early foetal life in red deer. *J. Embryol. Exp. Morphol.* **29,** 431–437.

Linsdale, J. M., and Tomich, P. Q. (1953). "A Herd of Mule Deer." Univ. of California Press, Berkeley.

Millais, J. G. (1906). "The Mammals of Great Britain and Ireland," Vol. 3, pp. 71–179. Longmans, Green, London.

Murie, O. J. (1951). "The Elk of North America." Stackpole Publ. Co., Harrisburg, Pennsylvania.

Peterson, R. L., ed. (1955). "North American Moose." Univ. of Toronto Press, Toronto.

Sempéré, A. J., and Lacroix, A. (1982). Temporal and seasonal relationships between LH, testosterone and antlers in fawn and adult male roe deer (*Capreolus capreolus* L.): A longitudinal study from birth to four years of age. *Acta Endocrinol. (Copenhagen)* **98,** 295–301.

Seton, E. T. (1929). "Lives of Game Animals," Vol. 3. Doubleday, Garden City, New York.

Sugden, L. B. (1964). An antlered calf moose. *J. Mammal.* **45,** 490.

Suttie, J. M. (1983). The influence of nutrition and photoperiod on the growth of antlers of young red deer. *In* "Antler Development in Cervidae" (R. D. Brown, ed.), Caesar Kleberg Wildl. Res. Inst., Texas A & I University, Kingsville (in press).

Tegner, H. (1961). Horn growth in infant roe deer. *Proc. Zool. Soc. London* **137,** 635–637.

Wika, M. (1980). On growth of reindeer antlers. *In* "Proceedings of the Second International Reindeer/Caribou Symposium" (E. Reimers, E. Gaare, and S. Skjenneberg, eds.), pp. 416–421. Direktoratet for vilt og ferskvannsfisk, Trondheim.

Zimmer, A. (1905). Die Entwicklung und Ausbildung des Rehgehorns, die Grösse und das Körpergewicht der Rehe. *Zool. Jahrb., Abt. Syst. (Dekol.), Geogr. Biol.* **22,** 1–58.

Developmental Anatomy
of Antlers

The eighteenth century scientist, Buffon, believed that deer antlers were made of wood. The French word for antler is *bois*, and the skin was once referred to as *écorce* (bark). Although the superficial similarities between the antlers of deer and the branches of trees are inescapable, antlers are of course bone. However, because the nature of antler bone has been the subject of so much debate over the years, they have appropriately been referred to as "bones of contention" (Henshaw, 1969).

Next to their regenerative abilities, the most unique attribute of deer antlers is the astounding rate at which they elongate. Bamboo shoots may grow faster, but nothing else in the animal kingdom, including cancer, surpasses the velocity of antler elongation. The evolution of such a prodigious developmental feat required that previous records of nerve and blood vessel growth be outdistanced, and that normal mechanisms of mineral metabolism in the body be exaggerated to mobilize the vast quantities of calcium and phosphorus deposited each year in growing antlers.

The elongation of an antler describes a typical **S**-shaped growth curve. It starts out slowly in the spring, accelerates exponentially during the summer,

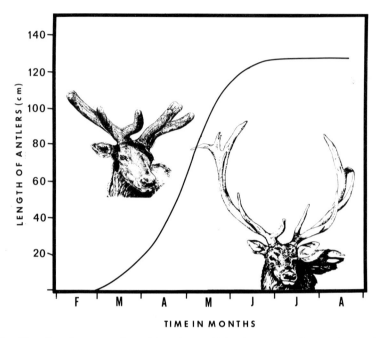

Fig. 73. Growth curve of wapiti antlers, which elongate at about 1.75 cm/day in midseason.

and slackens its growth as fall approaches (Fig. 73). The antlers of most species, large and small, remain in velvet for about the same period of 3–5 months, although the actual phase of elongation may approximate only 100 days. Large antlers, having farther to grow in the same period of time, must lengthen more rapidly than short ones. The antlers of a small species of deer, for example, measuring only 10 cm in length, add about 1 mm/day. Those of large species with meter-long antlers, augment their lengths an average of 1 cm/day. The 2-m antlers of the Irish elk must have had to add 2 cm/day. Since elongation of antlers is not linear, daily increments during the steepest inflection of the growth curve are almost double the overall means. If one combines the rates of elongation of the several tines growing simultaneously on both antlers, a deer may actually produce as much as 10 cm of new antler material every day in midseason. Van Ballenberghe (1983) has estimated that a bull moose in Alaska will generate 417 gm of tissue per day in June when his palmate antlers are growing maximally.

Rörig (1906, 1908) measured the patterns of branched antler growth in various species, including the red deer, wapiti, and roe deer. These analyses

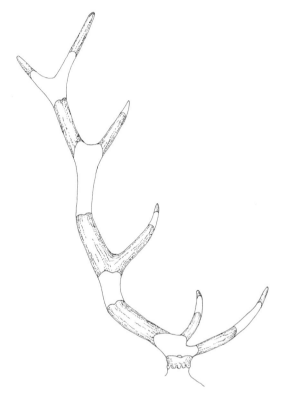

Fig. 74. Successive increments of red deer antler growth, comparing tines and main beam at intervals during the growing season. (After Rörig, 1906.)

confirm that the more proximal tines complete their development before the distal ones, and that successive stages of growth may be in progress simultaneously in different branches of a growing antler (Fig. 74). Comparable observations on caribou have been reported by Bergerud (1976).

The pattern of antler growth radiates distally from its center of origin (Fig. 75). Bifurcations elongate in divergent directions, the tines eventually decelerating while the main beam sustains the tempo of its upward thrust until the final dimensions are approached. In other expanding systems, intermittent pulses of growth leave their marks on the final product. The concentric striations that embellish mollusc shells and the alternating grooves and ridges engraved on the surfaces of horns, for example, bear witness to the physiological rhythms to which the growth process is so responsive. No

Fig. 75. Growth of antlers in a sika deer photographed on (a) May 7, (b) May 13, (c) June 5, (d) June 15, (e) August 23, and (f) January 6.

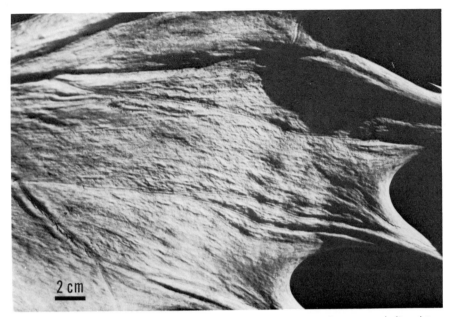

Fig. 76. Inside palm of a moose antler demonstrating concentric contours believed to represent pulses of daily growth.

such periodicities are evident on deer antlers, with one possible exception: moose antlers sometimes exhibit faint transverse ridges on the more distal regions of their palms, especially on the inner concave surface of the bone. If incident illumination is held at a low angle, the contours of these ossified undulations can be made visible (Fig. 76). Occurring at wavelengths of about 2 cm, these recurrent swellings may be tangible traces of a daily resurgence at the growing margins when the antler bone was being laid down.

Antler Casting

In the past it was believed that deer withdrew to secluded places to cast their antlers (Sobieski and Stuart, 1848), the implication being that the loss of virility was not something any self-respecting stag would wish to exhibit in public. Aristotle wrote that deer hid themselves because they were defenseless without their antlers. In "The Boke of Saint Albans," an early treatise on hunting, Dame Juliana Berners (1486) perpetuated the ancient superstition in a Middle English verse* about roe deer:

> At saynt andrew day his hornys he will cast
> In moore or in moos he hidyth hem fast
> So that no man may hem sone fynde
> Ellys in certayn he woos not his kynde

St. Andrew's day falls on November 30, as do the antlers of roe bucks. Yet such an instantaneous event as the casting of antlers is so seldom observed even by those who work closely with deer, that perhaps an understandable misconception is permissible. Indeed, the dearth of antlers found in nature must have appeared to substantiate the conviction that deer retreated to inaccessible localities to cast them. We now know that dropped antlers are a rich source of minerals avidly gnawed by other creatures, particularly rodents (Fig. 77). They may also be consumed by deer themselves. Reindeer and caribou have on occasion been observed to chew on each other's antlers, even before they have fallen off.

The onset of antler growth is in most cases closely tied to the casting of the old antlers. However, this relationship seems not necessarily to be one of cause and effect. In some species, the old antlers drop off several months before the new ones begin to grow. In most deer the two events coincide. Even in the latter instances, one often sees the swelling of the distal pedicle skin immediately beneath the burr, an event that forecasts the impending separation of the old antler from the pedicle. Even if the antler fails to fall off, as sometimes occurs under abnormal lighting conditions (Goss, 1969), or is prevented from doing so by artificially screwing the antler to the pedicle (Goss, 1963), antler regeneration can still occur from beneath or around the base of the still-attached antler (Chapter 9). Such growths cannot develop normally, of course, and are usually confined to a circular bulge around the base of the uncast antler, sometimes with branches sprouting from the basal ring of antler tissue (Behlen, 1906).

The foregoing evidence supports the contention that loss of the old and growth of the new are probably not causally related events. Each can occur in the absence of the other, but whether they are both triggered by the same physiological stimulus remains to be determined. The best estimate at pre-

---

*At saynt andrew day his hornys he will cast
In moore or in moos he hidyth hem fast
So that no man may hem sone fynde
Ellys in certayn he doos not his kynde

Fig. 77. Fallow antler showing where rodents have begun to gnaw around the edges.

sent is that declining concentrations of male sex hormone are responsible for these events, perhaps each being independently sensitive to different levels of testosterone.

The casting process has not received the attention it deserves. Presumably it is hormonally controlled because both antlers often drop off on the same day, sometimes within minutes of each other. Occasionally, one antler may be lost and the other carried for several weeks afterward, much to the inconvenience of the imbalanced buck. It was once believed that stags would dislodge their antlers by thrusting them into the mud (Sobieski and Stuart, 1848). There is one report (Newsom, 1937) that casting may be actively promoted in the moose by knocking the antlers against trees. However, Waldo and Wislocki (1951) observed that white-tailed deer "appear to be quite unaware of the impending event." The casting of antlers is probably as spontaneous as it is abrupt. Those who have worked with deer know that up to a few days prior to casting it is possible to drag the entire animal by his antlers, only to have them fall off of their own weight several days later.

The mechanism by which the dead antler is separated from the living pedicle is said to resemble the "abacterial sequestration" of necrotic bone in certain clinical situations in which aseptic ischemia of an appendage may eventually result in a form of self-amputation (Gruber, 1952a,b). Both of these processes are assumed to be mediated by the action of osteoclasts, multinucleate cells specialized for bone resorption. Although they are not known to have been specifically described in the case of the about-to-be-detached antler, they probably play a significant role in the erosion of bone at the base. Whether or not their action is stimulated by parathyroid hormone, as is the case in the resorptive process of normal bone turnover, is not yet known. As long ago as 1861, Wyman attributed antler casting to resorption of bone around the Haversian canals. This results in the separation of the old antler from the pedicle as the attenuated spicules of bone connecting the two are broken. The base of the cast antler, therefore, is rough in texture because of the protruding remnants of these spicules (Fig. 78). The base of the freshly cast antler is normally bloodless, although the pedicle from which it has been detached bleeds enough to produce a substantial scab (Fig. 79). One infers that the boundary along which osteoclasts align them-

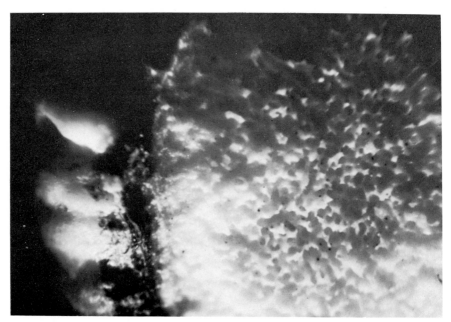

Fig. 78.   Base of cast antler with spicules of bone remaining after osteoclastic separation from the pedicle. (From Goss, 1963. Copyright 1963 by the American Association for the Advancement of Science.)

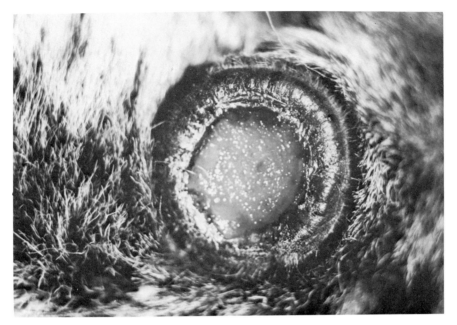

Fig. 79. Raw surface of a sika deer pedicle minutes after the old antler has been detached. The surrounding skin is already swollen as renewed growth is initiated.

selves must be immediately distal to the terminal blood vessels in the pedicle bone such that they are not disrupted until the antler has actually broken off.

In general, the casting of antlers differs from other kinds of autotomy in that it involves the loss of dead, not living, tissue. Nevertheless, the antler system conforms to the axiom that wherever a structure is capable of spontaneous autotomy, it is also capable of regeneration.

Histogenesis

The anatomy of antlers is best understood in terms of how they develop. They originate in the fawn or yearling from the periosteum (Chapter 6). The bone into which this tissue differentiates is responsible for inducing the overlying integument to become the velvet. It also gives rise to the bony pedicle which retains the potential to regenerate new antlers year after year. Identification of the specific tissues in the pedicle that give rise to the regenerated antler has been a challenging but elusive problem for the experimental zoologist. Attempts to pinpoint the histologic origin from which antlers regenerate each year have thus far yielded inconclusive results. The reason for this is, in part, the widespread distribution of antlerogenic tissues in the vicinity of the pedicle, and the inconvenient efficiency with which selectively removed tissues regenerate themselves, thus frustrating the intent of deletion experiments.

It would seem logical that a bony structure such as an antler should arise from the bony component of the pedicle. However much these two kinds of bones are bound together, this is not proof that one is derived from the other. Attempts to provide more convincing evidence have involved the surgical excision of the pedicle bone. In the most radical resections, the entire pedicle has been amputated (Jaczewski, 1955), or substantial amounts of surrounding frontal bone have also been chiseled out of the skull (Bubenik and Pavlansky, 1956). Although these operations may sometimes preclude antler production in the same year, such deer usually grow new antlers, often smaller than normal, the year after. Not only does the wound heal over with new skin, but bone is also regenerated in the places from which it has been removed. Following less heroic operations, antler regeneration is not delayed. If the pedicle skin is peeled back and the bone sawed off, new bone grows in the empty cylinder of pedicle skin, and antler formation ensues (Goss, 1961). When the distal one-half of the pedicle, bone and skin together, is amputated in the winter or early spring before the bony antlers have been lost, the cut surface remains unhealed until the time comes for casting and regrowth to occur, whereupon the wound heals belatedly and antler growth commences. Delayed wound healing under these circumstances is interesting. Wound healing, especially in the integument, is such a universal attribute of living systems that it is nearly impossible to prevent. Yet in the antler pedicle, healing appears to be held in abeyance until the right season of the year when physiological conditions conducive to the onset of antler regeneration prevail.

Except for blood vessels, nerves, and connective tissue, the antler pedicle is literally skin and bone. There are neither joints nor skeletal muscle. Although the antler epidermis is derived from that on the pedicle, it is not inconceivable that the dermis of the pedicle skin might give rise to the rest of the antler. This is difficult to prove by deletion experiments. It is possible to cut the skin away from the pedicle, but subsequent wound healing in the spring is responsible for restoring the pedicle skin and promoting antler growth (Goss, 1961). However, it is worth noting that such antlers are not necessarily normal. During their early stages, they may appear swollen, presumably owing to the accumulation of fluids (Bubenik et al., 1956). A similar reaction has been described by Mautz (1977) following ligation of the pedicle. This condition may reflect an imbalance between the arterial and venous blood flows as a result of the surgical disruption of pedicle vessels when the skin is removed (Chapter 9). Such edematous antler buds may be transient if the cause does not persist; a nearly normal antler eventually is produced.

Further attempts to explore the possible role of skin in antler production have utilized transplantation techniques. These may involve the grafting of antler or pedicle skin elsewhere on the body to see if it will give rise to antlers, or the transplantation of exogenous skin to the pedicle (Goss, 1972). In the former case, free grafts of pedicle skin discs to the ear have failed to give rise to antler tissue, although the grafts themselves survived and grew normal pedicle-type hairs (Fig. 80). Transplants of velvet skin from growing antlers to the deer's hind leg (Billingham et al., 1959) or scalp (Goss, 1972) have been found to survive for up to several years, during which period they preserved the typical appearance of antler skin (Fig. 81) but failed to give rise to antler outgrowths in following years. Curiously, these grafts did not die when the velvet of the antlers from which they were derived was shed. This is convincing evidence that the seasonal demise of antler skin is not innately caused by the programmed death of its cells, but must be attributed

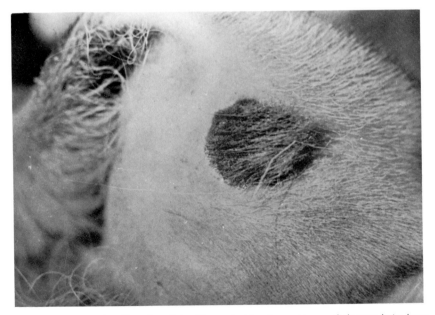

Fig. 80. Graft of a disc of pedicle skin on inside of ear. New pedicle-type hairs have regenerated in the absence of antler or velvet differentiation.

Fig. 81. Autotransplant of velvet skin from the growing antler of a sika deer to the scalp. As seen 1 year later, it has retained the relatively sparse pelage of its original derivation, but has not given rise to an antler. In the absence of subjacent bone, it has remained viable while the host's antlers have gone through their usual cycles of replacement.

to external agencies, presumably ischemic necrosis. The shedding of antler velvet, therefore, is a case of murder, not suicide.

In other experiments, it has been possible to replace some of the pedicle skin with that from another region of the body. Because the configuration of the pedicle does not lend itself to the successful application of free grafts, it is more convenient to perform flap grafts (Fig. 82) by taking advantage of the proximity of the ear to the pedicle (Goss, 1964). This is done by circumcising the distal 1 cm of pedicle skin in the late winter or early spring before the old antlers have been cast. Once the ring of pedicle skin has been removed, a hole is cut through the full thickness of the ear, which is then impaled on the pedicle. Naturally, it is first necessary to saw off most of the old antler so that the ear hole can be pulled over the pedicle. Once in position, the inner epidermis of the ear heals to the remaining epidermis of the pedicle. However, the outer ear epidermis, having no skin with which it may complete its continuity, is unable to heal its wound until such time as the old antler is cast. Once this occurs, the outer ear epidermis migrates over the raw end of the pedicle from which the antler has been detached, a process similar to

Fig. 82. Sequence of stages in which ear epidermis participates in antler development. (A) The perforated ear is pulled over the sawn off antler to fit around the circumcised pedicle. (B) After casting the old antler, the pedicle wound is healed with outer ear epidermis. (C) A normal antler bud develops, using ear skin as the only available source of epidermis. (D) The resulting antler is normal except for the absence of the brow tine. (From Goss, 1964. Reprinted with permission from Pergamon Press, Ltd.)

that normally achieved by the pedicle epidermis. Under these experimental conditions, the epidermis of the pedicle skin has no way to participate in antler development because it has already fused with the epidermis of the inner ear skin.

Despite the lack of participation on the part of the pedicle epidermis, normal antlers are capable of developing under these conditions. Such antlers are covered with what appears to be normal velvet skin bearing hairs that are definitely of antler type rather than ear type. The implications of this experiment are obvious. The differentiation of epidermis, including its glands and appendages, in the direction of antler velvet must be induced by underlying tissues. If something as different as ear epidermis can become velvet, presumably almost any type of epidermis could do likewise under comparable circumstances. Indeed, the demonstration that antlerogenic periosteum from fawns, when transplanted beneath leg skin, can give rise to pedicle and antler tissue (Fig. 71) confirms the plasticity of the epidermis in this regard. The mesodermal components of the antler covered with epider-

Fig. 83.   Red deer, photographed on July 25, bearing a third antler on his forehead derived from the autotransplantation of an antler bud. (Courtesy of Dr. Zbigniew Jaczewski.)

mis derived from the ear is undoubtedly of pedicle origin. Although an ear grafted to the end of the pedicle can interfere with the participation of pedicle epidermis in antler development, there is no reason to believe that it necessarily prevents the migration of mesodermal cells to the pedicle stump where they can presumably take part in antler development as they normally do. Therefore, this experiment sheds little light on the mesodermal histogenesis of antlers.

Grafts of entire antler buds have probably been attempted more often than reported in the literature. Free grafts of such magnitude, in the absence of vascular anastomoses, are subject to extensive necrosis. Notwithstanding these difficulties, successful transplants have been achieved by Jaczewski (1956a,b, 1958, 1961) in the red deer and fallow deer, and by Pavlansky and Bubenik (1960) in the roe deer. Autografts of early antler buds transplanted elsewhere onto the skull with or without parts of the adjacent pedicles, have been shown to be capable of elongating into antlers (Fig. 83). Even when substantial portions of the grafts become necrotic, sufficient amounts of tissue may survive to give rise to new growth. Although such antlers tend to be stunted, they shed their velvet, are cast, and regrow in synchrony with nongrafted ones.

It is significant that transplanted antlers can be regenerated in following years, even when no pedicle tissues have been included in the original graft. This suggests that cells from the growing antler that would otherwise be marked for destruction can survive in the graft site as seeds of future antler generations. The morphogenetic relationship between such subsequent outgrowths and the proximodistal level of the original graft from which the residual cells were derived is an interesting problem for further investigation.

Wound Healing

The earliest indications of antler growth are observed in the tumescence of the distal pedicle skin. This swollen ring of integument has a different appearance and texture from that which covers the rest of the pedicle. Its shiny surface is characteristic of the antler velvet into which it will develop. As soon as the old antlers have been cast, this skin migrates over the stump of the pedicle. It does so by the ingrowth of both epidermal and mesodermal cells. The former insinuate themselves between the dead and desiccated material of the overlying scab and the viable tissues beneath. The latter include the mesodermal cells migrating between the epidermis and the exposed bone of the pedicle. Histological sections of the healing process give the impression that cells from the dermis of the pedicle skin and of the periosteum surrounding the pedicle bone migrate into the healing area that is later to grow into the antler bud (Waldo and Wislocki, 1951; Wislocki, 1942). Although the static view seen in a histological preparation must be interpreted with caution, the participation of all stump tissues in the production of a regenerate might be inferred by extrapolation from other regenerating appendages. Judging from the histogenesis of the original antler in the fawn's frontal periosteum, the pedicle periosteum of the adult is an obvious candidate for the possible source of the bony component of the regenerating antler. It cannot be decided at this juncture whether or not dermis gives rise only to dermal tissues and periosteum only to bone. The original mass of undifferentiated and hyperplastic cells that compose the antler bud could be derived from any and all tissues in the pedicle stump. The true origins and destinies of these cells will not soon be deciphered with limited existing techniques of investigation.

The most intriguing aspect of antler bud formation is that the pedicle skin does not form a scar. Elsewhere on the body, comparable wounds heal by the prompt differentiation of a fibrous mass of connective tissue beneath the wound epidermis. Such scars are in fact regenerated dermis, but a dermis that lacks the orderly configuration of collagen fibers typically seen in the normal skin. When a structure regenerates, it produces a bud, or blastema, instead of a scar. For reasons not yet understood, this is what happens in the healing of pedicle stumps too. Indeed, injuries to the side of the pedicle have been reported on occasion to give rise to small, laterally directed, antlers (Nitsche, 1898). The skin of the pedicle would appear to be a very special kind of integument, endowed with the capacity to deflect its otherwise natural tendency for scar formation in the direction of antler bud production. To the extent that dermis contributes cells to the antler, the very tissue that elsewhere blocks regeneration by precocious scar formation may itself give rise to the mass of cells without which antler regeneration cannot proceed. In this perspective, the antler could be regarded as an exaggerated version of scar formation.

The importance of wound healing in the onset of antler regeneration cannot be overemphasized. Like virtually all other cases of epimorphic regeneration (Chapter 8), the replacement of antlers begins with the healing of the raw surface on the stump. Indeed, if the pedicle stump is sealed with full-thickness skin grafts (Goss, 1972), antler regeneration is prevented altogether because there is no wound from which it can originate.

The Nature of Velvet

The velvet of a growing antler is a special type of integument. It is capable of enormous expansion to keep pace with the rapid elongation of the tissues it envelopes and to which it is so closely adherent. The glands and appendages that adorn it must multiply profusely. It is richly vascularized and innervated, although specialized sensory endings are not apparent. The antler velvet is unlike any other kind of skin, if only because it is uniquely adapted to the promotion of mammalian appendage regeneration.

The hairs are particularly interesting. They are formed de novo at the growing tip of the antler, being left behind as the velvety pelage lining the shaft while elongation advances. In some deer (e.g., sika) the antler hairs are sparce. In others (e.g., wapiti, reindeer) they form a furry surface that obscures the underlying epidermis. In either case, the skin covering the growing tip constantly differentiates new hair follicles, a phenomenon conspicuous by its absence in most kinds of mammalian skin (Billingham et al., 1959). In longitudinal sections of the growth zone one can observe successive stages in the initiation and development of hair follicles (Fig. 84) as the section is examined from the apex to the shoulder of the growing tip. Not

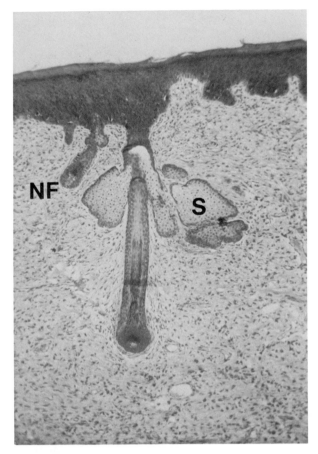

Fig. 84.   The hairs that adorn antler skin are associated with conspicuous sebaceous glands (S). At the growing tip, follicles differentiate *de novo* (NF). (From Goss, 1964. Reprinted with permission from Pergamon Press, Ltd.)

only are hair follicles differentiated, but each is accompanied by a sebaceous gland that secretes its oily substance into the shaft of the hair follicle when they are fully differentiated. It is this sebum that accounts for the shiny appearance of the velvet.

The fully developed hairs of the velvet are unusual in that they lack the arrector pili muscles associated with hair follicles elsewhere on the body. It is these little muscles that are responsible for the erection of hairs, and the production of "goose bumps." In the antler, the hairs tend to grow out at nearly right angles to the surface of the skin, a position responsible for the velvety texture of the antler pelage. Whether or not such hairs can regenerate when plucked, as is true with all other hairs on the body, has not been explored. However, it is an interesting point because antler velvet is a transient tissue, and its hairs do not exhibit the seasonal molting characteristic of the rest of the body surface. One could argue that to preserve their regenerative ability would be a useless redundancy, one which would seem to have had little or no selective advantage in the course of evolution. It would be interesting to learn the answer.

There is yet another enigma of antler hairs, one that is especially evident in the growing antlers of wapiti and fallow deer. The velvet of these antlers often exhibits a longitudinally oriented striped pattern not unlike the grain of wood (Fig. 85). The graceful arches of these contours suggest that a family of hair follicles sharing similar pigmentation might have been produced in a single line of descent from cells in a particular region of the original antler bud. The nature of the grain in the velvet has neither been investigated nor explained.

One final aspect of antler epidermis is worth mentioning. When the pedicle epidermis heals over the pedicle stump and thus envelopes the antler bud itself, there are long tongues of epidermis situated at the center which protrude down into the underlying proliferating mesodermal cells (Fig. 86).

Fig. 85.   The graceful curvatures in the "grain" of the velvet may be visible evidence of earlier growth patterns in the development of this fallow buck's antlers.

Fig. 86. In the early antler bud of the sika deer, elongate papillae in the apical epidermis extend deep into the underlying mesoderm.

Whether these configurations are actual downgrowths, or columns of epidermis left behind as the underlying cells push upward, is a relative question. Nevertheless, their existence is transient, being characteristic of the original antler bud, but not of the growing tips once elongation is well under way. These epidermal configurations are not unlike the epidermal thickenings and downgrowths encountered in other developing systems (Chapter 8). The limb bud of the embryo, for example, is equipped with an apical epidermal cap which plays an important role in the outgrowth and morphogenesis of the limb bud. The blastema of regenerating amphibian limbs forms beneath a thickened cap of epidermis that develops in the healing wound over the stump. In regeneration from the margins of holes punched through rabbit ears there are conspicuous epidermal downgrowths, transient in nature, adjacent to the site of blastema formation. Such structures have not been described in nonregenerating ears. Whether or not the conspicuous and curious tongues of epidermis that penetrate the apex of the antler bud relate to seemingly comparable structures in other developing systems cannot be determined; there is no categorical evidence that they play a significant role at all in the development of the antler. However, their very presence arouses curiosity.

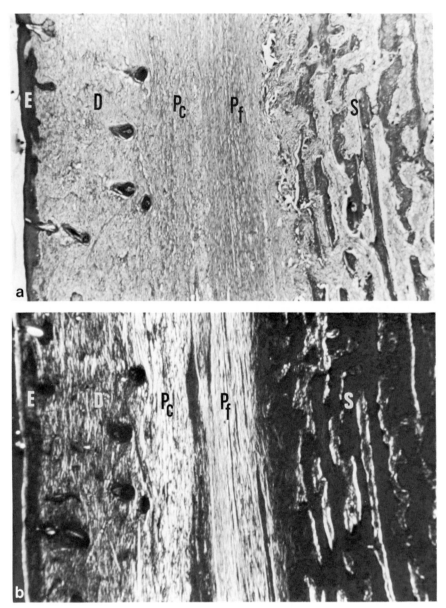

Fig. 87.   Photomicrographs of longitudinal sections from along the side of a growing white-tailed deer antler, as seen in hematoxylin-stained bright field (a) and unstained polarized light (b). E, epidermis; D, dermis; P_c, periosteum (showing coarse collagen fibers); P_f, periosteum (fine fibers); S, spongiosa. Under polarized light microscopy the orientation of collagen fibers may be visualized in relation to associated histologic structures. (Courtesy of Dr. Donald P. Speer.

As elsewhere in the body, the dermis consists of a thickened layer of collagen fibers. At the growing ends of the antler these fibers are less organized than lower down where they are predominantly longitudinal in orientation. Indeed, it is almost impossible to strip the skin off a mature antler except along its length in parallel with the dermal collagen fibers. Immediately beneath the dermis is the perichondrium, or periosteum, depending on the level. Speer (1983) has demonstrated that this layer is subdivided into an outer stratum populated by coarse collagen fibers and an inner one of fine fibers (Fig. 87). In the apical growth zone the longitudinal fibers are augmented by circumferential ones. Along the antler shaft, only the former are found.

Zones of Differentiation

During the course of antler development there are persistent growth zones at the tips of the branches. These are developmental descendants of the original antler bud and its subsequent bifurcations (Fig. 88). Their cells, reminiscent of fibroblasts, are in a high state of proliferation. They spin out quantities of collagen fibers, and will eventually differentiate into the various tissue components of the antler. However, as long as proliferation at the apex surpasses differentiation in subapical regions, elongation continues. Therefore, a longitudinal section through the end of a growing antler reveals in spatial array those events that are in fact separated in time (Fig. 89).

The growing tip of an antler may be divided into a number of zones representing successive stages in the differentiation of its cells. Delineation of the zones is indistinct, reflecting the continuous, not intermittent, differentiation of the cells and tissues involved. Beneath the envelope of epidermis and dermis lies a hyperplastic layer of perichondrium (Banks, 1974; Speer, 1983). Its outermost cells, referred to as "reserve mesenchyme" by Banks (1974) and Banks and Newbrey (1983a), show few if any signs of differentiation (Fig. 90). However, their deeper descendants are recognizable as chondroblasts, confirming the perichondral nature of the tissue in which they reside. This layer of perichondrium is continuous peripherally with the more proximal periosteum that surrounds the shaft of the antler immediately inside the dermis of the velvet. The perichondrium is richly endowed with collagen fibers, some encircling the apical growth zone, others running longitudinally (Speer, 1983). It is also highly vascularized by the cascade of vessels that drains blood from the growing tip (Fig. 91).

The cap of perichondrium gives rise to the cartilaginous zone, composed of trabeculae of chondrocytes and the matrix within which they lie (Fig. 92). Between these trabeculae are numerous vascular channels returning blood

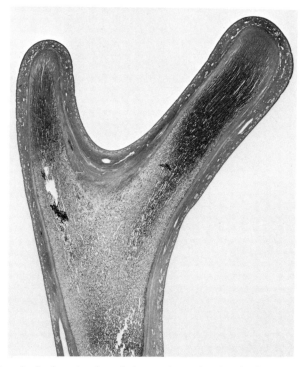

Fig. 88. Longitudinal section through the growing antler of a sika deer. Arteries are visible in the velvet, giving rise to innumerable small vessels converging downward at the growing tips. Below the bifurcation, the inner substance of the antler consists of spongy bone.

to the pedicle. The cartilage later becomes calcified before bone formation supervenes. The region of calcified cartilage, where the earliest stages of ossification become evident, has been designated the primary spongiosa by Banks (1974). Below this is the secondary spongiosa, composed of cancellous bone in the central locations and more compact bone peripherally (Banks, 1974). Progressive ossification leads to the formation of increasingly solid bone in more proximal regions, a process that creeps upward as the antler elongates and the growing season advances.

Chondrification

Antler growth is bone growth. As such, the mechanism by which ossification takes place has been the subject of considerable investigation, and the object of no small measure of controversy. Because the cells that proliferate

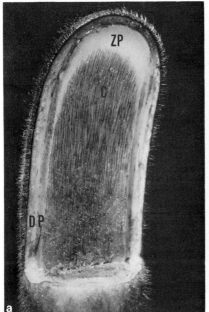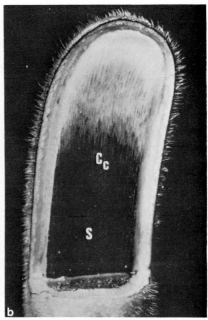

Fig. 89. Growing tine of a sika antler sliced open longitudinally. The fresh preparation (a) has its vascular channels rendered visible by their contained blood. The same specimen has been rephotographed (b) following von Kossa (AgNO₃) staining to reveal locations of calcium in black. C, cartilage; C_c, calcified cartilage; D, dermis; P, periosteum. S, spongiosa; ZP, zone of proliferation.

in the growing tip produce collagen, they should probably be classified as fibroblasts, at least during this phase of their lives. Slightly deeper ones sees the alignment of these cells into columns. Farther down, successive stages in their transformation can be traced through the perichondrium into the cartilaginous zone.

This is no ordinary cartilage. Unlike that elsewhere in the body, antler cartilage is honeycombed with vascular channels. It is arranged in vertical columns alternating with the blood vessels in between (Fig. 92). Yet every histologist knows that almost by definition, cartilage is not supposed to be vascularized. The fact that this rule is violated argues against the existence of cartilage in the growing antler. In point of fact, cartilage sometimes is vascularized. The articular cartilage in the epiphyses of bones in very large animals, such as the elephant, is said to contain vascular channels. This is an adaptation to the metabolic demands of chondrocytes embedded in a matrix far removed from the nearest source of supply. The high metabolic rate of growing antlers likewise demands a copious blood supply. Cartilage

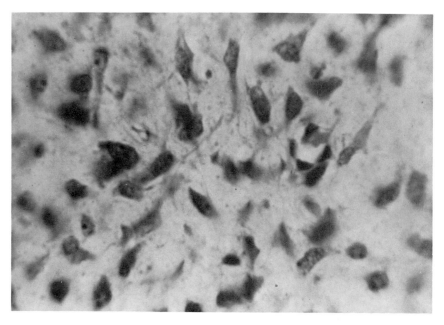

Fig. 90.   Undifferentiated cells in the proliferative zone of the growing antler tip. Hematoxylin and eosin. (From Goss, 1970.)

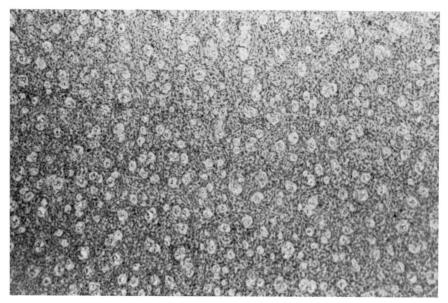

Fig. 91.   Cross-section through the procartilage at the antler apex, showing numerous blood vessels draining blood from the growing tip. Hematoxylin and eosin. (From Goss, 1970.)

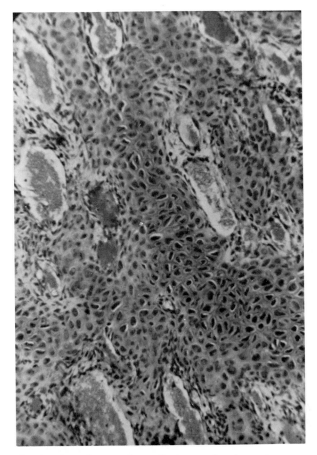

Fig. 92. In the chondrogenic zone, the cartilaginous trabeculae are interspersed with longitudinally oriented blood vessels. Hematoxylin and eosin. (From Goss, 1970.)

comes in many configurations, and it may be prudent not to define it within limits that are too narrow.

Nevertheless, there were some investigators who even denied the existence of chondrogenesis in antler development, pointing out that "The growing bone imperceptibly merges with the undifferentiated connective tissue" (Noback and Modell, 1930). The cartilagelike tissue was referred to as "preosseous tissue" by Modell and Noback (1931) and Wislocki et al. (1947), or as "preosseous mesenchyme" by Kuhlman et al. (1963). Eventually admitting the existence of cartilage per se, Bejsovec (1954) and Wislocki (1956) concluded that the chondrocytes were converted directly into osteocytes and the cartilaginous matrix into bone ground substance.

Molello et al. (1963) suggested that chondrocytes trapped in osteoid appeared to become osteocytes. The possibility that chondro-osteo-metaplasia may actually occur cannot be ruled out, but the current consensus favors endochondral ossification as the dominant mechanism by which antler bone is formed.

The zone of chondrification is different from the cartilaginous growth plate of other endochondral bones (Banks and Newbrey, 1983b). Its cells are not aligned in columns, although the tissue itself is organized into parallel trabeculae. In antlers, the cartilage is already vascularized and need not therefore be eroded by invading blood vessels as is the case in the cartilaginous plates of long bones, but the two systems have much in common. They are both examples of polarized growth along the longitudinal axis of the bone under construction. Each is histochemically similar, as demonstrated by the metachromatic staining of sulfated mucopolysaccharides, the presence of glycogen, and the existence of alkaline phosphatase activity (Frasier et al., 1975; Kuhlman et al., 1963; Lojda, 1956; Molello et al., 1963). In both, the chondrocytes tend to hypertrophy as the matrix becomes calcified (Sayegh et al., 1974).

Wislocki et al. (1947) called attention to the fibrous nature of the antler cartilage, and Newbrey and Banks (1975) described the ultrastructural details by which collagen fibers are synthesized in the perichondrium. Collagen persists as longitudinally oriented fibers in the cartilaginous trabeculae (Speer, 1983). However, this protein consists of types I and III collagen, in the absence of type II collagen that normally forms in the hyaline cartilage of mammals (Newbrey et al., 1983).

Ossification

Between the cartilaginous and osseous zones is a relatively narrow stratum where endochondral ossification occurs. The cartilaginous trabeculae become calcified and chondroclasts are visible on their surfaces (Banks, 1974; Banks and Newbrey, 1983a; Lodja, 1956). Considerable erosion of the cartilage takes place, although not all of the cartilaginous matrix is destroyed. Concomitant with chondroclasia, osteoblasts appear and begin to lay down bone matrix on the surfaces of the remnants of the trabeculae. The exact source of the osteoblasts is not known, although perivascular connective tissue has been suggested as a likely origin (Banks and Newbrey, 1983b; Modell and Noback, 1931; Molello et al., 1963). It is by the appositional deposition of bone that ossified trabeculae are formed (Fig. 93), replacing their cartilaginous precursors.

Once bone has differentiated, it occurs in the configuration of a network

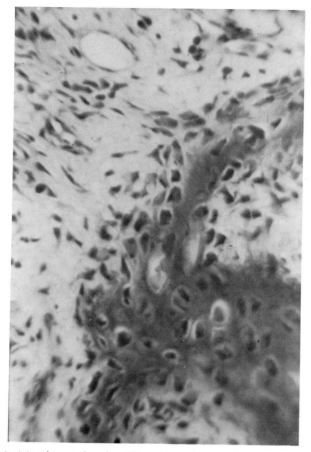

Fig. 93. Incipient bony trabeculae differentiate in the spongiosa. Hematoxylin and eosin. (From Goss, 1970.)

of trabeculae (Fig. 94). This spongy bone, interspersed with numerous blood vessels, comprises most of the shaft of the growing antler. Even in the peripheral regions, where in other skeletal elements one would expect to find circumferential lamellae of compact bone, only a labyrinth of bony trabeculae exists.

As final maturation of the fully grown antler takes place, the spongy bone in its shaft solidifies, more than doubling its density by the end of the growing season (Brown et al., 1978). This is achieved by appositional ossification, mediated by osteoblasts, on the surfaces of preexisting trabeculae. The resultant thickening of the trabeculae, at the expense of the vascular spaces in between, gradually curtails the flow of blood within the

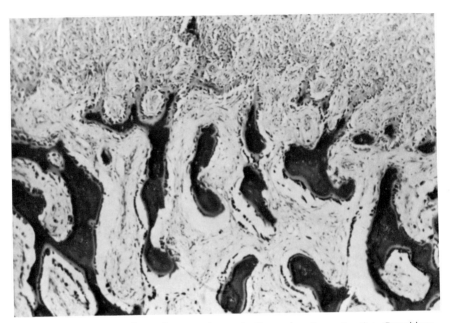

Fig. 94. Bony trabeculae at the apex of an antler tine undergoing maturation. Osteoblasts line the bone surfaces. Hematoxylin and eosin. (From Goss, 1970.)

antler. However, the innermost portions of the antler may remain more porous than the cortical areas, permitting a trickle of residual blood flow for a time, even after the velvet has been shed. Eventually, the substance of the antler becomes converted into almost solid bone, by which time the blood flow is fully shut off and the last basal regions of the antler die back to the level of the pedicle.

Antler bone is lined with periosteum, but of a type that is considerably thicker than that on ordinary bones. Because this periosteum gradually merges with the connective tissue and dermis on the outside, its exact limits are not easily defined. Nevertheless, it is in some respects not unlike the cambium of a tree. It is responsible for the development of "pearlation," the knobby appearance of the antler surface. In reindeer, injuries to the surface of the growing antlers may sometimes cause the outgrowth of extra tines, presumably derived from the traumatized periosteum (Bubenik, 1959).

However unique the antler bone and cartilage may be, their composition is not significantly different from that of other skeletal elements in the body. When in velvet, the cartilage of an antler possesses the usual array of enzymes (e.g., dehydrogenases, alkaline phosphatase) found in cartilage elsewhere (Banks and Mayberry, 1974). The growing tips incorporate phosphate, as indicated by ^{32}P uptake, maximally at these sites (Bernhard et al., 1953; Bruhin, 1953). Although ^{32}P is incorporated to a lesser extent in the peripheral regions of the shaft, there is little or no uptake in the pedicle. In the mature, bony antler after the velvet has been shed there is no ^{32}P uptake, confirming the metabolic inertness of the dead antler bone.

This does not mean that dead antler bone does not change with time. Chapman (1981) has noted that antlers tend to dry out after rut, and hunters know that the weights of trophy specimens are maximal in freshly killed deer. This probably accounts for the increasingly more brittle nature of antlers as the season progresses (Currey, 1979). Ash comprises only about 20% of the weight of velvet antlers, but well over 50% of that of bony ones (Hyvarinen et al., 1977; Ullrey, 1983). The mineral content of antlers is not significantly different from that of other bones, being composed primarily of calcium and phosphorus. Collagen is the chief protein.

Systemic Effects of Antler Ossification

The composition of the blood might be expected to reflect the metabolic activity in antlers, if only because this is the medium by which materials are transferred from the rest of the skeleton to the growing antler. Nevertheless, blood calcium and phosphorus levels have been found to remain relatively constant at all times of year, irrespective of whether the antlers are in velvet or bony (Bernhard et al., 1953; Graham et al., 1962). They have not been found to differ significantly between the sexes (Murphy, 1969). However, alkaline phosphatase, an enzyme associated with heightened metabolic activity, is three to five times higher in the serum of male fallow and white-tailed deer during the summer when they are in velvet, compared with the winter when the antlers are bony and not growing (Graham et al., 1962). This enzyme is also higher in the serum of fawns that in adult deer. In does, it remains constant all year. If the growing antlers are amputated or ligated at their bases, serum alkaline phosphatase levels decrease, a reaction not observed when only one antler is removed. It is possible, therefore, that the growing antlers may be the primary source of alkaline phosphatase in serum, an interpretation consistent with the findings that this enzyme occurs in high concentration in the cartilaginous tissue near the growing tips of antlers (Kuhlman et al., 1963).

The production of antlers, particularly their mineralization, is a tremendous drain on the system, especially in larger species possessing disproportionately large antlers. There are two ways that growing antlers could affect mineral balance in the body. One would be for the deer to consume extra quantities of salts during the period when the antlers are growing, thereby depositing materials directly into the developing antlers. Alternatively, such minerals might be incorporated into the other bones of the body, later to be withdrawn and reutilized in antler construction. Investigations have confirmed that the latter mechanism is the method by which minerals are mobilized for antler growth. For example, there is an increase in the rate of turnover in skeletal elements when deer are growing antlers, compared with the nongrowing season. This turnover is evident in the increased numbers of osteons being resorbed at any one time, particularly in those bones (e.g., ribs) that are not weight bearing (Banks et al., 1968a,b; Hillman et al., 1973; Meister, 1956). Thus, deer experience an osteoporotic reaction while growing their antlers. Studies with radioactive isotopes confirm that minerals in the diet are for the most part deposited in the body's skeleton before being resorbed and redeposited in the growing antler (Cowan et al., 1968). Therefore, deer apparently do not store excess minerals in their skeleton in anticipation of antler growth (Rerabek and Bubenik, 1956), but accelerate the turnover of such substances during the growing season.

As annually renewable structures, antlers have served a unique role in enabling man to monitor radioactive fallout, a problem that became a matter for serious concern in the 1950s when nuclear bombs were being tested by a number of nations in different parts of the world. Hawthorne and Duckworth (1958) analyzed radioactive strontium in the antlers of red deer on the island of Islay in Scotland and found a tenfold increase from 1952 to 1957. Gelbke (1972) reported a correlation between ^{90}Sr in red deer antlers and the emission of fission products from atmospheric nuclear explosions between 1947 and 1969. The ^{90}Sr in the jaw bones of black-tailed deer shot in California revealed about 30 times as much radioactivity in 1960 as in 1952 (Schultz and Longhurst, 1963). Because antlers do not accumulate strontium perennially as mandibles do, the existence of numerous trophies properly documented with the dates and places where they were shot by hunters is a valuable record from which past levels of environmental radioactivity can be monitored in the years to come.

Vascularization

A rapidly growing structure such as the antler requires lots of nourishment. This is provided by the copious flow of blood to the growing tips. The blood flow is so great that the surface temperature of growing antlers is noticeably warm to the touch. Deer antlers are probably the only external mammalian structures in which the temperature equals that of the deep body. Thermograms of reindeer clearly show the tips of the antlers as among the warmest parts of the body (Krog et al., 1969). So impressive is this effect that it has been suggested that antlers may have evolved to dissipate excess heat from the body (Stonehouse, 1968). It cannot be denied that antlers are in fact thermal radiators (Chapter 5). However, it is doubtful that their main function is to get rid of metabolic heat in males during the summer months when they are building themselves up for the autumn rutting season. Such a hypothesis is incompatible with the occurrence of antlers in both sexes of reindeer and caribou, and the winter growing season of the roe deer. When sika deer were exposed to wide swings of ambient temperatures, their antlers in velvet underwent corresponding fluctuations, although rectal temperatures remained constant (Ohtaishi and Too, 1974). However, in reindeer, antler temperatures failed to respond to stress of heat and cold (Wika et al., 1975). Thermoregulatory reactions may therefore be different from one species to another. The luxurious blood flow in deer antlers is most logically interpreted as a mechanism to ensure elevated temperatures conducive to the rapid proliferation of cells in a structure that must grow fast enough to reach full dimensions in just a few months.

In July of 1785, the English anatomist, John Hunter, carried out an ingenious experiment on a stag in Richmond Park outside London (Stevenson, 1948). This was no mean feat in the era before anesthesia when animals had to be forcibly restrained during such an operation. Hunter tied off the external carotid artery leading to the growing antler on one side of the head. He noted that the surface temperature of that antler dropped postoperatively, but that 1 week later it had again become warm as growth resumed. Subsequent autopsy revealed that smaller branches proximal to the ligature had enlarged enough to restore the normal blood flow to the affected antler. Hunter thus discovered the phenomenon of collateral circulation, a vascular reaction of considerable importance in surgical procedures that involve arterial ligations.

The superficial temporal artery (Fig. 95) supplies blood to the growing antler (Rörig, 1900; Waldo et al., 1949). A dozen or so major branches from this artery may grow into the developing antler from the pedicle. This cutaneous blood flow may be supplemented by vessels derived from the pedicle bone, at least during the early stages of antler growth. Later on, the importance of the latter may diminish along with that of the venous return through the vessels of the pedicle bone. If a ligature is tied tightly around the proximal part of a young growing antler early in the summer, the distal portion of the antler is not entirely cut off from its blood supply (Waldo et al., 1949). It may become edematous for about 1 week, but in due course it will resume its normal pattern of growth. This suggests that although the cutaneous arteries and veins may be constricted, there is sufficient afferent and efferent blood flow through the core of the growing antler to sustain it. The swelling of such ligated antlers is presumably attributable to the transient

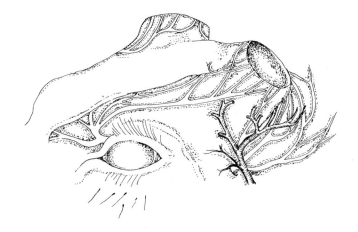

Fig. 95.   Anatomical arrangement of arteries (dark) and nerves (white) in the antler region of a white-tailed deer. (After Wislocki and Singer, 1946.)

interference with venous return. If a ligature is tied around a more mature antler later in the growing season, the distal portions become necrotic because so much of the blood flow during the later stages of antler growth is in the skin rather than internal. This shift in the arrangement of blood vessels in the growing antler reflects the increasing ossification of the antler shaft as maturation of the antler progresses, an ossification that is increasingly less compatible with the persistence of significant blood flow internally.

The prominence of arteries in the velvet is impressed upon the underlying bone of the antler. Examination of the surface of bony antlers reveals channels on the surface where the deposition of bone has been molded by arteries in the overlying velvet (Fig. 96). Although much of the blood carried by these arteries is delivered directly to the growing tips, some of it is diverted inward along the shaft via recurrent arterioles. These may terminate superficially in the underlying peripheral bone or penetrate to the inner regions of the antler (Rhumbler, 1929, 1931). However, at the apex all of the arterial blood is channeled vertically downward through the growth zone and delivered to medullary sinuses in the cartilage. Depending on the age of the antler, this blood may continue down through the core to the pedicle, or be returned through a venous collecting system to major veins in the velvet. There may occasionally be arteriovenous anastomoses in the velvet where arterial blood can be shunted directly into neighboring veins without first dividing into a capillary network.

The histology of antler arteries reveals their similarity to the arteries of the

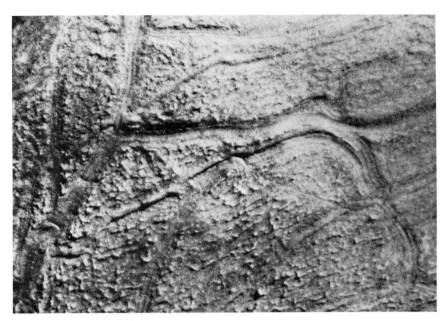

Fig. 96. Impressions of arteries, originally in the velvet, left on the underlying bone of a moose antler.

umbilical cord. Antlers and placentas share in common a transient life span. Both are vulnerable to the types of injuries that could rupture major arteries, bleeding from which could not be controlled by blood clotting. Accordingly, antler arteries and umbilical arteries are adapted to constrict in response to local stimuli (Altura et al., 1972; Wika and Krog, 1980). Perhaps this unusual attribute is shared by the arteries of amphibian and reptilian tails capable of autotomy. If a growing antler is sawed off, blood will spurt several meters from the severed arteries. However, within minutes, one can observe the slackening of flow until bleeding has been effectively stopped by constriction at the level of amputation.

Antler arteries have unusually thick walls and relatively narrow lumens (Waldo et al., 1949). Unlike arteries elsewhere in the body, their histology is not conveniently divided into the tunica intima (the innermost sheath of connective tissue), tunica media (a layer of smooth muscle fibers responsible for the contractility of arteries), and tunica adventitia (an outer sheath of fibrous connective tissue). Although these three tissues are present in antler arteries, they are not clearly delineated by the internal and external elastic membranes normally separating the media from the intima and adventitia, respectively (Wika and Edvinsson, 1978; Wika and Krog, 1980). The layer of smooth muscle fibers is clearly adapted for prompt constriction and their action can be studied in isolated antler arteries (Wika and Krog, 1980). When immersed in saline solution they are observed to undergo spontaneous and rhythmic peristaltic contractions. Whether or not these arteries are themselves directly innervated by antler nerves remains to be confirmed (Wika and Edvinsson, 1978; Vacek, 1955). However, like other arteries of the body, they have been shown by Wika and Edvinsson (1978) to contract *in vitro* in response to perfusion with various biogenic amines (e.g., adrenaline, histamine). Therefore, antler arteries are responsive to chemical, mechanical, and perhaps neural stimuli.

Innervation

Antlers are richly innervated. Their sensitivity is made clear to anyone who attempts to touch antlers in velvet, which usually elicits a vigorous avoidance reaction on the part of the deer. There is presumably also a kinesthetic sense, whereby deer seem to be aware of where the points of their antlers are, even when not visible. It is this sense that enables such large species of deer as wapiti, moose, and woodland caribou, to trot through forests without hitting their growing antlers on the branches of trees. Presumably the memory of antler shape persists even after the velvet and nerves have been lost, judging from the skill with which bony antlers are manipulated in threshing and combat.

Direct observations by Wislocki and Singer (1946) of the nerves in antlers reveal that they derive from the supraoptic and temporal branches of the trigeminal nerve (Fig. 95). Both myelinated and nonmyelinated nerve fibers occur in antlers (Vacek, 1955). They tend to grow out along with the major arteries in the velvet.

In nonantler skin, nerves are especially dense in the arrector pili muscles of the hair follicles, muscles which are not present in the velvet. Nerves are also concentrated in sweat glands, sebaceous glands, and blood vessels. In the antlers, nerves are most abundant in the hair follicles, as well as in the intervening epidermis (Vacek, 1955). There are no specific sensory end organs, such as Pacinian corpuscles. Although antler arteries (Wika and Edvinsson, 1978), like umbilical arteries (Lachenmeyer, 1971), have been claimed not be be innervated, unmyelinated nerves have been found by Vacek (1955) in the tunica media of arteries, and myelinated nerves in the adventitia.

Innervation of the growing antler occurs by the regeneration of pedicle nerves. These nerves are distinguished by their uncommonly thickened epineurium, the connective tissue sheath which envelops them (Waldo et al., 1949). Unlike the nerves of amputated mammalian appendages, which tend to form neuromas from the tangle of regenerating fibers with nowhere to go, pedicle nerves remain unregenerated for two-thirds of the year between the shedding of the velvet and the casting of the antlers. When the time comes, they are capable of regenerating at the same rapid rate at which the antler grows.

There has been one published account of the effects of antler denervation. This was reported by Wislocki and Singer (1946) who had reason to suspect that antler regeneration, like that of appendages in lower vertebrates, might depend on an adequate nerve supply. Accordingly, they cut the nerves leading to the right antler pedicles on two white-tailed deer in mid-March, after the old antlers had been cast but before the new ones had begun to grow. However, despite denervation these antlers began to grow at the normal time in April, but failed to attain the typical shape and dimensions of the intact left antlers. The abnormalities that occurred were attributed not so much to the lack of neurotrophic influences per se as to the secondary effects of desensitization. Unable to know where their denervated and insensitive antlers were, the deer could not avoid injuries. Repeated trauma to such antlers was probably responsible for their developmental abnormalities. Therefore, although nerves are in fact indispensable for the regeneration of certain appendages in other vertebrates (Chapter 8), they seem not to play a comparable role in antler regeneration. One might speculate that innervation is necessary for regeneration only when it is required for the function of the regenerated structure. Antlers lack muscle and movement, as well as direct sensory perceptions in the fully mature structure. Therefore, it is perhaps not surprising that their regeneration is independent of the nerve supply.

Nevertheless, it has been proposed that certain aspects of antler development may in fact be subject to neurotrophic influences. On rare occasions, for example, antlers are encountered which have shed the velvet from the distal portions but not the proximal regions (Bubenik, 1966; Bubenik and Pavlansky, 1965). Because such a differential reaction is difficult to explain in terms of hormonal influences, it is conceivable that it might be mediated by nerves. It has also been shown that trauma to the growing antler may promote the production of extraordinarily enlarged antlers (Bubenik et al., 1982). When these are eventually cast off they may be replaced the following year by similarly abnormal structures. Again, this could be interpreted in terms of the effects of nerves on antler growth, and the ability of such nerves, or of neurons in the central nervous system with which they are connected, to "remember" stimuli from previous years. Bubenik and Pavlansky (1965) describe other cases in which abnormalities caused by injury to the growing antler were repeated in successive years. Like the inheritance of acquired characteristics, such anomalies challenge us to explain how morphogenetic information can be communicated from one generation of antlers to the next.

References

Altura, B. M., Malaviya, D., Reich, C. F., and Orkin, L. R. (1972). Effects of vasoactive agents on isolated human umbilical arteries and veins. *Am. J. Physiol.* **222,** 345–355.

Banks, W. J. (1974). The ossification process of the developing antler in the white-tailed deer (*Odocoileus virginianus*). *Calcif. Tissue Res.* **14,** 257–274.

Banks, W. J., and Mayberry, L. F. (1974). Selected enzyme histochemistry of antler ossification. *Anat. Rec.* **178,** 303.

Banks, W. J., and Newbrey, J. W. (1983a). Light microscopic studies of the ossification process in developing antlers. *In* "Antler Development in Cervidae" (R. D. Brown, ed.). Caesar Kelberg Wildl. Res. Inst., Texas A & I University, Kingsville (in press).

Banks, W. J., and Newbrey, J. W. (1983b). Antler development as a unique modification of mammalian endochondral ossification. *In* "Antler Development in Cervidae" (R. D. Brown, ed.). Caesar Kleberg Wildl. Res. Inst., Texas A & I University, Kingsville (in press).

Banks, W. J., Jr., Epling, G. P., Kainer, A., and Davis, R. W. (1968a). Antler growth and osteoporosis. I. Morphological and morphometric changes in the costal compacta during the antler growth cycle. *Anat. Rec.* **162,** 387–398.

Banks, W. J., Jr., Epling, G. P., Kainer, A., and Davis, R. W. (1968b). Antler growth and osteoporosis. II. Gravimetric and chemical changes in the costal compacta during the antler growth cycle. *Anat. Rec.* **162,** 399–406.

Behlen, H. (1906). Zur Gehörnentwicklung des Rehbocks im besonderen und der Cerviden im allgemeinen. *Zool. Beobachter* **47,** 262–269, 289–297.

Bejsovec, J. (1954). Ossifikace stitné chrupavky srnce (*Capreolus capreolus capreolus*). *Cesk. Morfol.* **2,** 169–178.

Bergerud, A. T. (1976). The annual antler cycle in Newfoundland caribou. *Can. Field Nat.* **90,** 449–463.

Berners, J. (1486). "The Boke of Saint Albans." Edit. Wynkyn de Worde.

Bernhard, K., Brubacher, G., Hediger, H., and Bruhin, H. (1953). Untersuchungen über chemische Zusammensetzung und Aufbau des Hirschgeweihs. *Experientia* **9,** 138–140.

Billingham, R. E., Mangold, R., and Silvers, W. K. (1959). The neogenesis of skin in the antlers of deer. *Ann. N.Y. Acad. Sci.* **83,** 491–498.

Brown, R. D., Cowan, R. L., and Griel, L. C. (1978). Correlation between antler and long bone mass and circulating androgens in white-tailed deer (*Odocoileus virginianus*). *Am. J. Vet. Res.* **39,** 1053–1056.

Bruhin, H. (1953). Zur Biologie der Stirnaufsätze bei Hufteiren, Teil I. *Physiol. Comp. Oecol.* **3,** 63–127.

Bubenik, A. B. (1959). Ein weiterer Beitrag zu den Besonderheiten der Geweihtrophik beim Ren. *Z. Jagdwiss.* **5**(2), 51–55.

Bubenik, A. B. (1966). "Das Geweih." Parey, Hamburg.

Bubenik, A., and Pavlansky, R. (1956). Von welchem Gewebe geht der eigentliche Reitz zur Geweihentwicklung aus? II. Mitteilung: Operative Eingriffe auf den Rosenstöcken der Rehböcke, Capreolus capreolus (Linné, 1758). *Saugetierkd. Mitt.* **4,** 97–103.

Bubenik, A. B., and Pavlansky, R. (1965). Trophic response to trauma in growing antlers. *J. Exp. Zool.* **159**, 289–302.

Bubenik, A., Pavlansky, R., and Rerabek, J. (1956). Trophik der Geweihbildung. *Z. Jagdwiss.* **2**, 136–141.

Bubenik, G. A., Bubenik, A. B., Stevens, E. D., and Binnington, A. G. (1982). The effect of neurogenic stimulation on the development and growth of bony tissues. *J. Exp. Zool.* **219**, 205–216.

Chapman, D. I. (1981). Antler structure and function—a hypothesis. *J. Biomech.* **14**, 195–197.

Cowan, R. L., Hartsook, E. W., and Whelan, J. B. (1968). Calcium-strontium metabolism in white-tailed deer as related to age and antler growth. *Proc. Soc. Exp. Biol. Med.* **129**, 733–737.

Currey, J. D. (1979). Mechanical properties of bone tissue with greatly differing functions. *J. Biomech.* **12**, 313–319.

Frasier, M. B., Banks, W. J., and Newbrey, J. W. (1975). Characterization of developing antler cartilage matrix. I. Selected histochemical and enzymatic assessment. *Calcif. Tissue Res.* **17**, 273–288.

Gelbke, W. (1972). Radioaktivität in Rotwild-Stangen und menschliche Strontium-90-Aufnahme. *Z. Tierphysiol., Tierernähr. Futtermittelkd.* **29**, 178–195.

Goss, R. J. (1961). Experimental investigations of morphogenesis in the growing antler. *J. Embryol. Exp. Morphol.* **9**, 342–354.

Goss, R. J. (1963). The deciduous nature of deer antlers. *In* "Mechanisms of Hard Tissue Destruction" (R. Sognnaes, ed.), Publ. No. 75, pp. 339–369. Am. Assoc. Adv. Sci., Washington, D.C.

Goss, R. J. (1964). The role of skin in antler regeneration. *Adv. Biol. Skin* **5**, 194–207.

Goss, R. J. (1969). Photoperiodic control of antler cycles in deer. II. Alterations in amplitude. *J. Exp. Zool.* **171**, 223–234.

Goss, R. J. (1970). Problems of antlerogenesis. *Clin. Orthop. Relat. Res.* **69**, 227–238.

Goss, R. J. (1972). Wound healing and antler regeneration. *In* "Epidermal Wound Healing" (H. I. Maibach and D. T. Rovee, eds.), pp. 219–228. Year Book Med. Publ., Chicago.

Graham, E. A., Rainey, R., Kuhlman, R. E., Houghton, E. H., and Moyer, C. A. (1962). Biochemical investigations of deer antler growth. Part I. Alterations of deer blood chemistry resulting from antlerogenesis. *J. Bone Jt. Surg., Am. Vol.* **44A**, 482–488.

Gruber, G. B. (1952a). Über das Wesen den Cerviden-Geweihe. *Dtsch. Tierärztl. Wochenschr.* **59**, 225–228, 241–243.

Gruber, G. (1952b). Studienergebnisse am Geweih des *Cervus capreolus*. *Zentralbl. Allg. Pathol. Pathol. Anat.* **88**, 336–345.

Hawthorn, J., and Duckworth, R. B. (1958). Fall-out radioactivity in a deer's antlers. *Nature (London)* **182**, 1294.

Henshaw, J. (1969). Antlers—The bones of contention. *Nature (London)* **224**, 1036–1037.

Hillman, J. R., Davis, R. W., and Abdelbaki, Y. Z. (1973). Cyclic bone remodeling in deer. *Calcif. Tissue Res.* **12**, 323–330.

Hyvärinen, H., Kay, R. N. B., and Hamilton, W. J. (1977). Variation in the weight, specific gravity and composition of the antlers of red deer (*Cervus elaphus* L.). *Br. J. Nutr.* **38**, 301–311.

Jaczewski, Z. (1955). Regeneration of antlers in red deer, *Cervus elaphus* L. *Bull. Acad. Pol. Sci., Ser. 2* **3**, 273–278.

Jaczewski, Z. (1956a). Free transplantation of antler in red deer (Cervus elaphus L.). *Bull. Acad. Pol. Sci., Cl. 2* **4**, 107–110.

Jaczewski, Z. (1956b). Further observations on transplantation of antler in red deer (Cervus elaphus L.). *Bull. Acad. Pol. Sci., Cl. 2* **4**, 289–291.

Jaczewski, Z. (1958). Free transplantation and regeneration of antlers in fallow deer [*Cervus Dama* (L.)]. *Bull. Acad. Pol. Sci.* **6,** 179–182.

Jaczewski, Z. (1961). Observations on the regeneration and transplantation of antlers in deer Cervidae. *Folia Biol.* (*Krakow*) **9,** 47–99.

Krog, J., Reite, O. B., and Fjellheim, P. (1969). Vasomotor responses in the growing antlers of the reindeer, *Rangifer tarandus. Nature* (*London*) **223,** 99–100.

Kuhlman, R. E., Rainey, R., and O'Neill, R. (1963). Biochemical investigations of deer antler growth. II. Quantitative microchemical changes associated with antler bone formation. *J. Bone Jt. Surg., Am. Vol.* **45A,** 345–350.

Lachenmayer, L. (1971). Adrenergic innervation of the umbilical vessels. Light and fluorescence microscopic studies. *Z. Zellforsch. Mikrosk. Anat.* **120,** 120–136.

Lojda, Z. (1956). Histogenesa parohů našich Cervidů a její histochemický obraz. *Cesk. Morfol.* **4,** 43–62.

Mautz, W. W. (1977). Control of antler growth in captive deer. *J. Wildl. Manage.* **41,** 594–595.

Meister, W. W. (1956). Changes in histological structure of the long bones of white-tailed deer (*Odocoileus virginianus*) during the growth of the antlers. *Anat. Rec.* **124,** 709–721.

Modell, W., and Noback, C. V. (1931). Histogenesis of bone in the growing antler of the Cervidae. *Am. J. Anat.* **49,** 65–86.

Molello, J. A., Epling, G. P., and Davis, R. W. (1963). Histochemistry of the deer antler. *Am. J. Vet. Res.* **24,** 573–579.

Murphy, B. D. (1969). The relationships of selected physiological factors to antler growth in mule deer. M. S. Thesis, Colorado State University, Fort Collins.

Newbrey, J. W., and Banks, W. J. (1975). Characterization of developing antler cartilage matrix. II. An ultrastructural study. *Calif. Tissue Res.* **17,** 289–302.

Newbrey, J. W., Counts, D. F., Foreyt, W. J., and Laegrejd, W. W. (1983). Isolation of collagen by guanidine extraction and pepsin digestion from the growing deer antler. *In* "Antler Development in Cervidae" (R. D. Brown, ed.). Caesar Kleberg Wildl. Res. Inst., Texas A & I University, Kingsville (in press).

Newsom, W. M. (1937). Winter notes on the moose. *J. Mammal.* **18,** 347–349.

Nitsche, H. (1898). "Studien über Hirsche." Englemann, Leipzig.

Noback, C. V., and Modell, W. (1930). Direct bone formation in the antler tines of two of the American Cervidae, Virginia deer (Odocoileus virginianus) and wapiti (Cervus canadensis). With an introduction on the gross structure of antlers. *Zoologica* (*N.Y.*) **11,** 19–60.

Ohtaishi, N., and Too, K. (1974). The possible thermoregulatory function and its character of the velvety antlers in the Japanese deer (*Cervus nippon*). *J. Mammal. Soc. Jpn.* **6,** 1–11.

Pavlansky, R., and Bubenik, A. (1960). Von welchem Gewebe geht der eigentliche Reiz zur Geweihentwicklung aus? IV. Mitteilung: Versuch mit Auto- und Homotransplantation des Geweihzapfens. *Saugetierkd. Mitt.* **8,** 32–37.

Rerabek, J., and Bubenik, A. (1956). Untersuchungen des Mineralstoffwechsels bei Geweihträgern mittels radioaktiver Isotopen. *Z. Jagdwiss.* **2,** 119–123.

Rhumbler, L. (1929). Injektionspraparate der Arterienwirbel an den Wachstumsenden von Hirschkolbengeweihen. *Verh. Dtsch. Zool. Ges.* **33,** 67–72.

Rhumbler, L. (1931). Ergänzende Mitteilungen über den Aderverlauf im Kolbengeweih der Hirsche, an Hand einiger Diapositive. *Verh. Dtsch. Zool. Ges.* **34,** 171–178.

Rörig, A. (1900). Über Geweihentwickelung und Geweihbildung. II. Abschnitt. Die Geweihentwickelung in histologischer und histogenetischer Hinsicht. *Arch. Entwicklungs mech. Org.* **10,** 618–644.

Rörig, A. (1906). Das Wachstum des Geweihes von Cervus elaphus, Cervus barbarus und Cervus canadensis. *Arch. Entwicklungs mech. Org.* **20,** 507–536.

Rörig, A. (1908). Das Wachstum des Geweihes von Capreolus vulgaris. *Arch. Entwicklungs mech. Org.* **25,** 423–430.

Sayegh, F. S., Solomon, G. C., and Davis, R. W. (1974). Ultrastructure of intracellular mineralization in the deer's antler. *Clin. Orthop. Relat. Res.* **99**, 267–284.

Schultz, V., and Longhurst, W. M. (1963). Accumulation of strontium-90 in yearling Columbian blacktailed deer, 1950–1960. *In* "Radioecology" (V. Schultz and A. W. Klement, Jr., eds.), pp. 73–76. Van Nostrand Reinhold, Princeton, New Jersey.

Sobieski, J., and Stuart, C. E. (1848). "Lays of the Deer Forest." William Blackwood & Sons, Edinburgh.

Speer, D. P. (1983). The collagenous architecture of antler velvet. *In* "Antler Development in Cervidae" (R. D. Brown, ed.). Caesar Kleberg Wildl. Res. Inst., Texas A & I University, Kingsville (in press).

Stevenson, L. G. (1948). The stag of Richmond Park: A note on John Hunter's most famous animal experiment. *Bull. Hist. Med.* **22**, 467–475.

Stonehouse, B. (1968). Growing antlers may have a thermoregulatory function. *Nature (London)* **218**, 870–871.

Ullrey, D. (1983). Nutrition and antler development in white-tailed deer. *In* "Antler Development in Cervidae" (R. D. Brown, ed.). Caesar Kleberg Wildl. Res. Inst., Texas A & I University, Kingsville (in press).

Vacek, Z. (1955). Innervace lýčí rostoucía parohů u Cervidů. *Cesk. Morfol.* **3**, 249–264.

Van Ballenberghe, V. (1983). Growth and development of moose antlers in Alaska. *In* "Antler Development in Cervidae" (R. D. Brown, ed.). Caesar Kelberg Wildl. Res. Inst., Texas A & I University, Kingsville (in press).

Waldo, C. M., and Wislocki, G. B. (1951). Observations on the shedding of the antlers of Virginia deer (Odocoileus virginianus borealis). *Am. J. Anat.* **88**, 351–396.

Waldo, C. M., Wislocki, G. B., and Fawcett, D. W. (1949). Observations on the blood supply of growing antlers. *Am. J. Anat.* **84**, 27–61.

Wika, M., and Edvinsson, L. (1978). *In vitro* studies of antler blood vessels. *Acta Physiol. Scand.* **102**, 70A.

Wika, M., and Krog, J. (1980). Antler "disposable vascular bed." *In* "Proceedings of the Second International Reindeer/Caribou Symposium" (E. Reimers, E. Gaare, and S. Skjenneberg, eds.), pp. 422–424. Direktoratet for vilt og ferskvannsfisk, Trondheim.

Wika, M., Krog, J., Fjellheim, P., Blix, A., and Rasmussen, U. (1975). Heat loss from growing antlers of reindeer (Rangifer tarandus) during heat and cold stress. *Norw. J. Zool.* **23**, 93–95.

Wislocki, G. B. (1942). Studies on the growth of deer antlers. I. On the structure and histogenesis of the antlers of the Virginia deer (Odocoileus virginianus borealis). *Am. J. Anat.* **71**, 371–416.

Wislocki, G. B. (1956). The growth cycle of deer antlers. *Ciba Found. Colloq. Ageing* **2**, 176–187.

Wislocki, G. B., and Singer, M. (1946). The occurrence and function of nerves in the growing antlers of deer. *J. Comp. Neurol.* **85**, 1–19.

Wislocki, G. B., Weatherford, H. L., and Singer, M. (1947). Osteogenesis of antlers investigated by histological and histochemical methods. *Anat. Rec.* **99**, 265–296.

Wyman, J. (1861). Account of some observations on the shedding of the antlers of the American red deer. *Proc. Boston Soc. Nat. Hist.* **7**, 167–168.

Regeneration

The most fundamental attribute of living things is that they can repair themselves. This regenerative ability derives from a basic capacity for growth and development inherent in all organisms. Reproduction itself is a form of regeneration whereby adults produce eggs that develop into replicas of themselves. However, true regeneration is the replacement of missing parts in adult organisms. The annual regrowth of antlers is a special case of this general phenomenon.

Modes of Growth

There are different means of replacing lost parts. In the normal wear and tear of everyday existence, there must be turnover of bodily components. This so-called physiological regeneration is responsible for the lifelong renewal of worn-out parts, a rejuvenating process responsible for counteracting the ceaseless depreciation of our bodies to which the phenomenon of aging is probably attributable. Physiological regeneration is most important at the molecular level as compounds are synthesized to replace those that are constantly degraded. The ultrastructural organelles into which these molecules are built may also turn over. Some specialized tissues of the body, such as nerves and muscles, must live as long as the organism. Others are designed to be renewed. Blood cells, for example, are manufactured in the marrow and lymphatic organs, live out a prescribed life span in the circulation, and then die as new ones replace them. The epithelial lining of the stomach and intestines is likewise maintained by the proliferation of new cells as fast as the old ones are sloughed into the gut. The epidermis is also subject to incessant turnover involving cell division in the basal layer commensurate with cell loss at the surface. In all renewing tissues there is a precise coordination between the rates of loss and replacement that maintains the tissue in a steady state. A slight imbalance between these rates of destruction and construction can result in such pathological consequences as dystrophy or cancer, depending on which way the scales tip.

Some parts of the body are capable of being replaced at levels of organization higher than that of the cell. In the ovaries, for example, there is a periodic turnover of follicles as new generations of eggs are produced with each reproductive cycle. Hairs and feathers, and sometimes even teeth, are regrown as their predecessors are shed. Other histologically complex structures reveal their capacities for regeneration only when accidentally lost or diminished. Such mechanisms of replacement are usually classified as tissue repair or wound healing (McMinn, 1969). Skin is the most vulnerable tissue in the body, and its capacities for wound healing are well known. When its continuity is interrupted, the epidermal cells on the surface migrate in from all directions to resurface the lesion. Concomitantly, the underlying dermis, a thick mat of collagen fibers, repairs itself by the synthesis of new connective tissue beneath the healing epidermis. Although the end result is not necessarily an exact duplicate of the original (new hair follicles are usually not produced), there is functional recovery even if the structural reconstitution is less than perfect. Internal tissues can likewise repair themselves. Broken bones undergo fracture healing by the development of a callus of considerable dimensions that is capable of differentiating into new bone to bridge the gap. Severed tendons spin out new collagen fibers to reunite their separated ends. Injured muscles are capable of regenerating to a limited extent in an effort to reestablish continuity across a lesion. Even visceral organs can repair themselves by a combination of cell division, migration, and redifferentiation. Indeed, there is no organ or tissue in the body without some capacity for self-repair, but such processes are little more than stopgap measures that remind us of how limited our own regenerative abilities really are.

Nevertheless, most organs and tissues can compensate for the inability to regenerate by a phenomenon known as compensatory growth (Goss, 1978). This is a process by which functional competence is established not by regrowing missing parts, but by enlarging what remains. Although morphological regeneration may not be achieved, functional recovery, which is more important, may be remarkably complete. Compensatory growth is best illustrated in the case of the liver which, when reduced by as much as two-thirds, undergoes a burst of cellular proliferation that reconstitutes the original mass of the organ in a matter of weeks. The shape of the restored liver is not the same as the original, but its functional competence is reestablished because the number of functional units is restored. Indeed, there is reason to believe that the compensatory growth of the liver is triggered by the functional insufficiency attending partial hepatectomy, and that growth is later turned off when physiological normalcy returns.

Most other organs in the body are also capable of compensatory growth, albeit to a less remarkable extent than in the case of the liver. If one kidney is removed (and even up to two-thirds of the other), that which remains undergoes considerable enlargement to make up for the deficiency. Loss of one lung likewise triggers expansion of the remaining one which in due course grows large enough to fill the chest cavity. Take out one adrenal gland and the opposite one nearly doubles its size. In organ after organ, one can demonstrate the latent capacity to grow larger in response to subtotal surgical resections or functional deficiencies. However, as convenient as such phenomena are, they are seldom as good as the regeneration of missing parts *in situ* would have been.

The Spectrum of Regeneration

Virtually all animals are competent to undergo physiological regeneration, wound healing, and compensatory growth of internal organs and tissues. However, where replacement of missing morphological components is concerned, there are vast differences in regenerative capacities among different types of creatures. Man and his fellow mammals are singularly lacking in such abilities. To appreciate our own regenerative incompetence one must view it in the perspective of how regeneration is distributed in the animal kingdom (Goss, 1969).

In general, it is well known that lower forms replace missing parts better than more highly evolved ones. Indeed, the lowest forms of life are even able to regenerate entire segments of their bodies. Single-celled animals, if cut in two, will replace the missing half if its nucleus remains intact. Multicellular organisms such as sponges, hydra, flatworms, and segmented worms possess the ability to become two organisms when cut in two, each part growing back that which has been lost. Such remarkable feats of regeneration have intrigued zoologists for over two centuries. Although these systems are very useful models in which to analyze basic problems of development, the mystery of how they can grow back such substantial fractions of their bodies remains to be solved.

As higher forms evolved from their wormlike ancestors, increased structural and functional complexities made it impossible to survive bisection, let alone regenerate missing heads and tails. Regeneration became limited to certain outgrowths from the body, the loss of which was not fatal. Thus, the starfish and octopus can grow back missing arms, and many insects can replace appendages, at least in larval stages. Crustaceans are especially well-endowed with regenerative abilities (Fig. 97), as the regrowth of lobster claws, for example, testifies. Many crustaceans have evolved special mechanisms for actively dropping off an injured leg. This spontaneous amputation, or "autotomy," is a useful adaptation for survival. Except for the discarded wings of ants and termites, and the detached placentae of mammals, autotomy is invariably accompanied by a highly developed capacity for regeneration. In a survey of several hundred green crabs in the intertidal zone, up to 30% of the individuals were found to have lost one or more legs and to be in various stages of replacing them at any given time. Clearly, in the rough and tumble conditions under which some organisms live, natural

Fig. 97. Regenerating lobster claw following autotomy of the original one at its base. The new appendage grows to considerable dimensions, but remains soft and immobile until the next molt.

selection seems to have favored the replacement of missing parts rather than mechanisms of avoiding their loss in the first place.

Regenerative abilities have persisted in the evolution of vertebrates. Fishes can grow back amputated fins and plucked scales. The taste barbels, or whiskers of the catfish, can regenerate too, including the numerous taste buds that pepper their surfaces. Salamanders can regenerate limbs (Fig. 98) and tails, but frogs and toads gradually lose the capacity to grow back their amputated legs during the course of metamorphosis (Wallace, 1981). This phenomenon, whereby regeneration is extinguished as a function of age and stage of development, has been used by some researchers as a model system in which to explore the mechanisms presumably responsible for the abolition of regeneration in higher vertebrates during the course of evolution. Thus, it has been shown that if extra nerves are deviated into the forelimb of a frog leg at a stage when it would normally have lost the capacity for regeneration, amputation leads to the regrowth of substantial portions of the missing arms at the end of the stump (Singer, 1951). These findings, coupled with the established fact that without nerves the regeneration of arms is impossible (Singer, 1974), have led to the hypothesis that the inability of lost appendages in warm-blooded mammals to grow back may be attributed to insufficient nerve supply. Unhappily, experiments designed to test this idea have not yielded the hoped-for results.

Reptiles are somewhat intermediate between amphibians and warm-blooded vertebrates, and their limited powers of regeneration reflect their level of evolution. Although lizards cannot regenerate legs, many can grow back amputated tails (Fig. 99) something which alligators and snakes cannot do. Indeed, as in the case of the discarded crab legs, regeneration accompanies their capacity for autotomy, having evolved as a strategy for survival. Anyone who has attempted to grab a lizard by the tail learns firsthand how

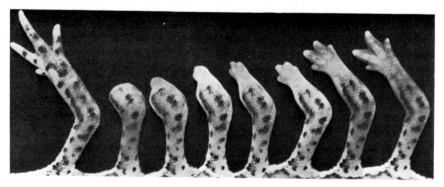

Fig. 98. Montage of successive stages in the regeneration of a newt limb 7, 21, 25, 28, 32, 42 and 70 days (left to right) after amputation. (From Goss, 1969.)

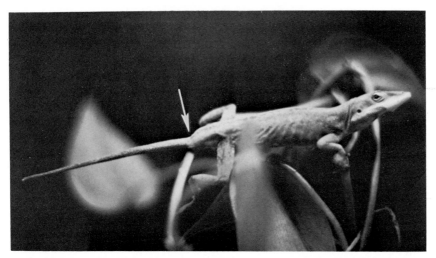

Fig. 99. An American "chameleon" (*Anolis carolinensis*) with a long regenerated tail produced following autotomy of the original at the level of the arrow.

readily its autotomy reflex can be triggered. Convulsive contractions of segmental musculature break off the tail at a preformed cleavage plane, and cause the lost part to wiggle for several minutes afterward. Remarkably little bleeding occurs at the stumps of such tails, presumably owing to the contraction of vascular musculature to constrict the blood vessels. The resemblance to the blood vessels in deer antlers is no coincidence. Regeneration of new tails proceeds rapidly, giving rise only to partial replicas of the original. Instead of being composed of segmented vertebrae, regenerated lizard tails possess a tapering unsegmented tube of cartilage extending from the last vertebra in the stump and containing an equally imperfect extension of the original spinal cord. Such hypomorphic tail regenerates in lizards contain segmented musculature, and if reamputated are themselves capable of repeated rounds of regeneration. Like the lost parts of the original tail, they are also able to twist and turn after being severed, thus holding the attention of predators while the lizard makes good its escape.

Mechanism of Blastema Production

Most of the foregoing examples are classified as epimorphic regeneration. That is, they involve the development of new structures to replace *in situ* those lost by amputation. This type of regeneration is achieved at the expense of cells surviving immediately proximal to the level of amputation.

They first undergo a process of dedifferentiation whereby the specialized attributes characteristic of their histological types are lost. Then they migrate distally to accumulate between the layer of wound epidermis and the severed portions of bone, muscle, and other mesodermal tissues immediately beneath the wound. The aggregation of these dedifferentiated cells at the end of the stump leads to the production of the blastema, a rounded mass of cells endowed with the capacity to develop into a structure replacing that which was lost (Fig. 100). Cells in the more proximal regions develop into the counterparts of muscle and bone in the stump. The growing tip of the blastema remains in a less differentiated state compatible with the need to proliferate new cells. Regeneration proceeds in a proximodistal direction, developing each part in sequence until the end of the appendage is reached. Apical proliferation ceases at this time and all cells differentiate as the process of regeneration completes itself.

There are certain features of the regenerative process that are shared in common by various appendages capable of replacing lost parts. These attributes tend to be the more important aspects of the process. Foremost among them is that something must be left behind from which new growth can proceed. Without a source of blastema cells, and without the necessary morphogenetic information, no meaningful regeneration can occur. Each appendage capable of regenerating missing parts consists of a "regeneration territory" that defines the boundaries within which cells capable of participation in the regrowth of that structure reside. If the entire territory is extirpated, then no opportunity for regrowth remains.

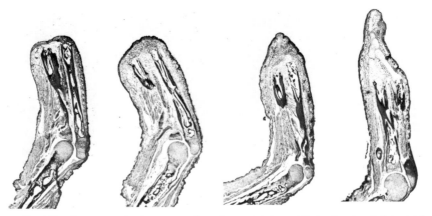

Fig. 100.   Four stages in the regeneration of amputated newt limbs showing (left to right) epidermal wound healing over stump, blastema formation, proximal differentiation of new cartilage, and morphogenesis. (From Goss, 1969.)

Another necessary ingredient of the regeneration process is the healing of an epidermal wound. Appendages may be defined as outgrowths from the body enveloped in skin. Unless this skin is interrupted, they will not regenerate. Amputation creates such a wound, and epidermal wound healing is one of the early events that initiates the replacement process. Indeed, this is what distinguishes epimorphic regeneration from tissue regeneration. Injuries to individual tissues (e.g., muscle, bone, tendon) may be followed by considerable local repair, but in the absence of a healing epidermal wound no blastema is formed and repair is restricted to the tissues that have sustained the injuries. In the case of epimorphic regeneration, more than just those tissues injured at the level of amputation are repaired. Epimorphic regeneration is not merely an extension of wound healing, differing from the latter in degree rather than type, but is a qualitatively distinct developmental phenomenon subject to a set of rules different from those that apply to tissue repair and wound healing. So important is epidermal wound healing to the onset of regeneration that blastema formation never occurs in its absence. One can imagine the evolution of this important relationship as a mechanism to ensure that appendages regenerate only when amputated. Otherwise, what would prevent limbs and tails from sprouting accessory outgrowths at random along their lengths?

A third requirement for epimorphic regeneration is a source of cells from which the blastema can arise. The histogenesis of regenerates has been a matter of conjecture and debate for many years. However, in appendages there is reason to suspect that most if not all mesodermal tissues in the stump contribute cells to the blastema. It does not necessarily follow that each cell type redifferentiates into the same type from which it was derived, although available evidence indicates that this is the usual course of events. Nevertheless, the possibility remains that such cells may also possess the capacity to change types (metaplasia) during their new reincarnation. It is very probable that connective tissue cells might become cartilage, or vice versa. However, a switch between cartilage and muscle fibers, for example, may surpass the metaplastic capacities of the cells involved. Further, epidermal cells do not become blastema cells proper, and nerve fibers are not derived from anything but themselves. The sources of new sheath cells to envelope these regenerated nerve fibers, and of regenerated blood vessels, remain to be determined.

In many examples of epimorphic regeneration it has been demonstrated that in order for the blastema to develop there must be an adequate supply of nerve fibers in the stump. This has been shown to be the case in amphibian limbs, the fins of fishes, and the taste barbels of the catfish. Probably the spinal cord in the tails of lizards and salamanders fulfills the same role. The nature of such neurotrophic influences is not clearly understood despite many years of intensive investigation. Presumably nerves are necessary in order to ensure that energy not be wasted in the regeneration of paralyzed structures. The role of nerves in certain cases of epimorphic regeneration constitutes a utilitarian imperative linking the process of growth to the demand for the functional replacement of lost parts. In the majority of regenerating structures, one can identify such physiological requirements, be they neural or hormonal, which emanate from sources outside the regenerating structures themselves but which exert important and indispensable influences on the initiation of appendage regeneration.

It is not enough just to produce a blastema and expect it to develop on its own. Its cells must contain information necessary to direct their future course of differentiation. Otherwise, morphogenetic chaos would ensue. The source of this information is either brought into the blastema by the participating cells, or is imposed on it by the remaining undifferentiated tissues in the stump. Either way, the blastema almost invariably develops into a remarkably flawless replica of the original appendage, a replica of the right size and the correct orientation. Such developmental events cannot be left to chance because a malformed appendage may be more useless than none at all. Equally impressive is the fact that regeneration replaces only those structures normally lying distal to the level of amputation. Even if a limb is grafted back onto the body in reversed polarity, it regenerates in the proximodistal direction despite the fact that it is being produced backward with respect to the tissues in the stump. Thus, blastema cells are imprinted with sufficient information to "know" from what level in the appendage they are derived, what the axes of orientation are, and what the dimensions of the regenerate must be in order to conform to those of the stump. Whatever may be the mechanism by which such morphogenetic information is communicated to the blastema cells, the latter read out their instructions and cease development when they reach the end of the message. Comparable mechanisms are undoubtedly responsible for determining the shapes and sizes of deer antlers when they are replaced each year.

Epimorphic Regeneration in Mammals

Not so many years ago there was little or no reason to believe that mammals could regenerate lost appendages. Not that biologists did not attempt to promote the regrowth of amputated parts, albeit unsuccessfully. Many experimental zoologists reluctantly concluded that there must be something about being a bird or a mammal that is incompatible with being able to regenerate, that the cells of higher vertebrates must be constitutionally incapable of participating in the process. Such a pessimistic view is no longer justified.

An alternative interpretation holds that the cells of birds and mammals may retain a latent capacity for regeneration but that their potentials along these lines are either prevented from being expressed by lack of some essential stimulus or are blocked by one or more inhibitory influences. This view is justified by the realization that mammals are not altogether devoid of the capacity for epimorphic regeneration. The existence of certain exceptions to the rule that they are not supposed to regenerate lends hope to the notion that, by appropriate experimental interventions, even structures normally incapable of regeneration might somehow be induced to grow back after amputation. Although the realization of this goal may be many years away, the relatively neglected subject of mammalian regeneration deserves more scientific attention than it has thus far received. It may be premature to attempt to promote regrowth in nonregenerative mammalian appendages, but the time has come to learn as much as possible about how regenerative ones can in fact replace their missing parts.

This is one reason the antlers of deer are so valuable to the developmental biologist. They are a case of natural regeneration in which appendages are spontaneously lost (like crab legs and lizard tails) and replaced each year by outgrowths from the stumps. The deer antler is a histologically complex appendage derived from a mass of undifferentiated cells that fits the description of a blastema. As such, it fulfills the definition of epimorphic regeneration. The very existence of these zoological curiosities is a challenge to explore why they evolved, what function they serve, and how they can regenerate. The amputation of other structures in deer, such as legs, tails, or ears, results in scar formation in the absence of regeneration. Wound healing occurs following all of these cases of amputation, but in the replacement of antlers the production of a scar is deflected into the developmentally more constructive pathways of blastema formation. In this context, it is altogether fitting that the prospects for regeneration in other mammalian structures be compared with what occurs in antlerogenesis.

A major breakthrough in the field of mammalian regeneration occurred in the Soviet Union in the early 1950s. Markelova, in her doctoral dissertation, explored the process by which rabbit ears regenerate new tissue to replace excised parts. This work was later described in a book on regeneration published in Moscow by her mentors, Vorontsova and Liosner (1960). Independent of this original discovery, the phenomenon was rediscovered by Joseph and Dyson (1966) in London. They were using the rabbit ear as a convenient site in which to study epidermal migration in healing skin wounds. In this system the skin is closely adherent to the underlying cartilage, thereby practically eliminating wound contraction. At the end of an experiment full thickness biopsies were taken for microscopic examination. When such rabbits were examined later on, it was found that the holes cut in their ears had grown in from the margins and had reestablished the continuity of the ear tissue.

Transversely amputated ears of rabbits do not grow back. Regeneration only occurs from the concave margins of holes, presumably because convergent outgrowths focus the growth zone into progressively smaller areas, thus maintaining a critical mass of proliferating cells to sustain regeneration. If a notch is cut in the side of an ear, regeneration occurs from the acute angle, but not from the straight edges. Holes 1 cm in diameter are readily filled in after 6–8 weeks (Fig. 101). Larger ones take longer, and often do not regenerate to completion.

When a hole is cut through the ear, the inner and outer epidermis migrate toward each other across the exposed margins (Goss and Grimes, 1972, 1975). In cross section, the ear consists of a sheet of cartilage sandwiched between two layers of skin. Nerves, blood vessels, and connective tissue are also present. The migrating epidermis insinuates itself between the desic-

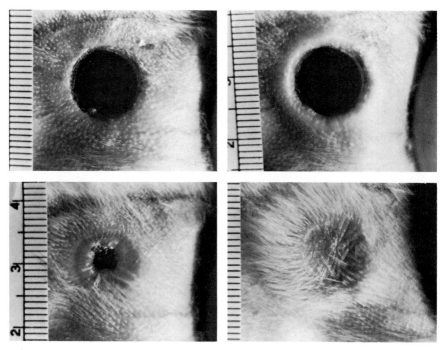

Fig. 101. Stages in the regeneration of tissues from the margins of a 1-cm hole cut through the full thickness of a rabbit ear. Photographed 1 day, and 1, 4, and 8 weeks postoperatively.

cated tissues at the wound margin and the underlying viable cells. It may take almost 1 week for the two sheets of migrating epidermis to make contact. In doing so, they usually grow through the cartilaginous sheet, the exposed margin of which dries out and is lost with the scab. Even before the migrating epidermis has sealed the wound, some of the cells migrate deep into the underlying tissues next to the cut edge of the dermis (Goss and Grimes, 1975). These epidermal downgrowths (Fig. 102) originate from both the inner and outer corners of the ear wound where their innermost cells subsequently undergo keratinization. This converts what was originally a solid core of epidermal cells into an infolding which, by the end of the second week, is pulled out flat from the tension generated by the expanding blastema.

While wound healing goes on, cells of uncertain mesodermal origin migrate toward the lesion where they accumulate between the wound epidermis and the severed margin of the cartilaginous sheet. Here they proliferate extensively and mediate the synthesis of collagen fibers. As the blastema rounds up and begins to elongate centripetally, some of the cells closest to the cut edge of the cartilaginous sheet undergo chondrogenesis. Thus, new

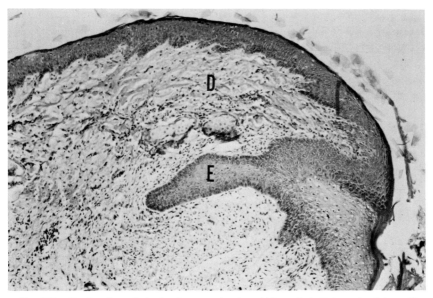

Fig. 102. Section through the healing margin of a rabbit ear hole 7 days after injury. The migrating epidermis (E) has penetrated deeply into the underlying tissues adjacent to the interrupted dermis (D). Hematoxylin and eosin. Epidermal downgrowths of this kind are evident from about 5–12 days postoperatively, eventually disappearing as the blastema forms. They are not found in nonregenerating ears.

cartilage differentiates off the end of old. Meanwhile, the growing tip of the blastema, which in a circular wound is actually doughnut shaped, continues to grow inward from the margins while the cells it leaves behind differentiate into a new cartilaginous sheet. In due course, the original aperture is reduced to a pinhole and then obliterated as the converging circular growth zone fuses in the middle. In this way, the continuity of the ear is reestablished, including its sheet of cartilage. Even occasional hair follicles are produced in the regenerated area.

It may be asked why the ears of the rabbit are so proficient in regenerating new tissue while those of other animals are incapable of doing so. A wide variety of mammals has been surveyed for their ability, or inability, to regenerate in this way, the vast majority of which are nonregenerative. These include the mouse, rat, guinea pig, chinchilla, hamster, gerbil, opossum, armadillo, Patagonian cavy (a rabbitlike rodent), dog, sheep, and deer. When sheep and dog ears are compared with those of rabbits, no epidermal downgrowths are found to occur in the healing wounds of the former's nonregenerating ears. They form scars around the margins of the holes instead of developing blastemas (Goss and Grimes, 1975). One wonders, therefore, if the epidermal downgrowths observed in rabbit ears may be causally related to the capacity of such ears to regenerate.

The roles played by component tissues in rabbit ear regeneration have been investigated by deletion and transplantation experiments (Goss and Grimes, 1972; Grimes, 1974a,b). There are two tissues in the ear which lend themselves to surgical manipulation, the cartilage and the skin. It is possible to remove much of the ear cartilage by peeling back a flap of skin resecting the cartilage sheet and replacing the skin as a flap graft. After healing, a hole can be punched through the region of the ear deprived of cartilage to learn if it can still regenerate. It does not (Fig. 103). This means that the cartilage plays an indispensable role in regeneration, although the exact nature of that role remains a mystery.

In other experiments, skin on either side of the ear has been removed and replaced with autografts of skin from the belly region of the rabbit (Goss and Grimes, 1972). A hole was then cut through the center of the operated region so that it was surrounded by belly skin. Only limited ingrowth took place under these conditions, and such regenerates as were formed lacked cartilage. Thus, the ear skin per se would appear to possess qualities conducive to normal regeneration.

It is tempting to conclude from the foregoing experimental results that the healing wound epidermis, or more particularly its epidermal downgrowths, may interact with the severed sheet of cartilage to set up conditions that permit regeneration to proceed. Regeneration of the full thickness of the ear can occur neither in the absence of a healing epidermal wound nor without

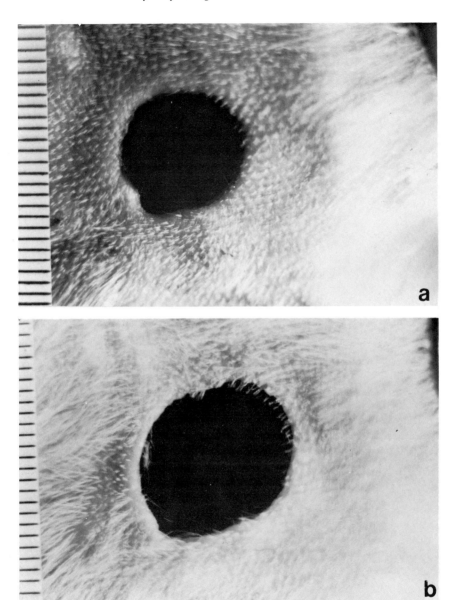

Fig. 103. Hole cut through a region of a rabbit ear previously deprived of its cartilaginous sheet. (a) 1 day after operation. (b) after 10 weeks no regeneration has occurred in the absence of cartilage participation.

the cartilaginous sheet. Cartilage is likewise unable to regenerate by itself unless its cut edge is adjacent to a healing wound in the integument. Both cartilage and epidermis seem to possess something the other lacks, such that only by interacting in close proximity are they able to cooperate in the complicated process of blastema formation and regeneration. Indeed, the importance of epidermal downgrowths is emphasized by the histological examination of the edges of holes cut through ears experimentally prevented from regenerating. As already mentioned, regeneration is effectively inhibited by removal of the cartilaginous sheet from the edges of such holes. This does not preclude wound healing by epidermal migration, but downgrowths from the epidermis fail to occur. Further, if the ear of the rabbit is exposed to high doses of X rays to prevent its regeneration, epidermal downgrowths likewise fail to develop. Both of these procedures abolish regeneration. Such circumstantial evidence does not prove that epidermal downgrowths have anything to do with regeneration, but they appear guilty by association.

It would seem that the rabbit ear is rather special. In relation to body size, it is the largest ear of any mammal. It is well adapted to pick up sounds, especially at night in open spaces where rabbits graze. However, rabbits are not the only animals that depend on their ears more than their eyes. Deer also possess large ears adapted to life out in the open, yet they have not evolved the capacity to regenerate missing tissue in them.

Rabbits are lagomorphs, a small order of mammals that includes hares and pikas. Hares have long ears like rabbits, and are equally capable of regeneration. However, pikas are short-eared lagomorphs that inhabit mountainous rock slides at high altitudes. They are famous for their habits of curing hay during the summer months and storing it for the long winters when the high passes are snowed in. Pikas are not easily kept in captivity, nor are they commercially available for experimentation. Consequently, if the regenerative abilities of their ears were to be tested it was necessary to capture wild specimens, punch holes in their ears, and release them for subsequent recapture. This in fact has been done,* and thanks to the territorial nature of these animals about two-thirds of the operated pikas were recovered some 2 months later. When their ears were examined it became clear that although the holes were not completely filled in, there was considerable ingrowth from around the margins. Microscopic examination confirmed that these regenerates also contained newly formed cartilage (Goss,

---

*The author is indebted to Dr. Preston Somers and his student, Carolyn Engel, of Fort Lewis College, Durango, Colorado, for their cooperation in supplying the pika ear tissues for this investigation.

1980). Thus, it would appear that there is something special about lagomorphs that enables them to regenerate ear tissue. What the selective advantages of this may be remains a matter for conjecture.

In the continuing quest for other mammalian ears capable of regeneration, it would seem logical to concentrate on those animals in which the ears play an especially important role. Bats suggest themselves as likely candidates. Accordingly, experiments have been conducted using flying foxes (*Pteropus*) (Goss, 1980, 1981). Holes were cut into their external ears, but no regeneration was observed. When it was realized that these bats do not fly by echo location, but by night vision, comparable studies were carried out on insectivorous bats (*Myotis*) and several species of Central American fruit bats that also fly by echo location. In all of these specimens, ear regeneration took place. Further, it was discovered that this type of regeneration is different from that which had been described in the rabbit. Although the holes punched through the ears of these bats were obliterated by marginal ingrowth, the tissues filling the gap consisted only of the two layers of skin and intervening connective tissue. No new cartilage differentiated in the regenerates. This may be taken to indicate that the capacity to regenerate ear tissue has evolved independently in different groups of mammals, and has therefore not always followed the same rules.

One other mammal has thus far been found capable of regenerating ear tissue following perforation. Like the rabbit, the domestic cat regenerates new tissue, including cartilage, to fill in holes. Prior removal of the cartilage prevents this regeneration. What is so special about cats that enables them to regenerate is difficult to fathom. However, the discovery that diverse orders of mammals are endowed with regenerative abilities lends hope to the possibility that other forms may share this trait. Comparisons of the mechanisms by which independently evolved processes of regeneration occur may yield useful information in understanding why regeneration takes place in some structures but not in others (Goss, 1981).

Ears and antlers are not the only things that regenerate in mammals (Goss, 1980, 1981). Bat wing membranes are vulnerable to being punctured or ripped, and are capable of regenerating new webbing (Fig. 104) to repair such injuries (Church and Warren, 1968). If a hole is punched in the eardrum of the guinea pig, the margins grow in to reestablish continuity (McMinn, 1975). The nictitating membrane of the pigeon's eye is likewise capable of regenerating new tissue if perforated. In newborn guinea pigs, cartilage of the nasal septum can be excised, following which a new septum is regenerated, provided the animal is less than 1 week old (Kvinnsland and Breistein, 1973). However, regeneration of holes punched through the nose leaf of Central American fruit bats, the combs or wattles of roosters, or the

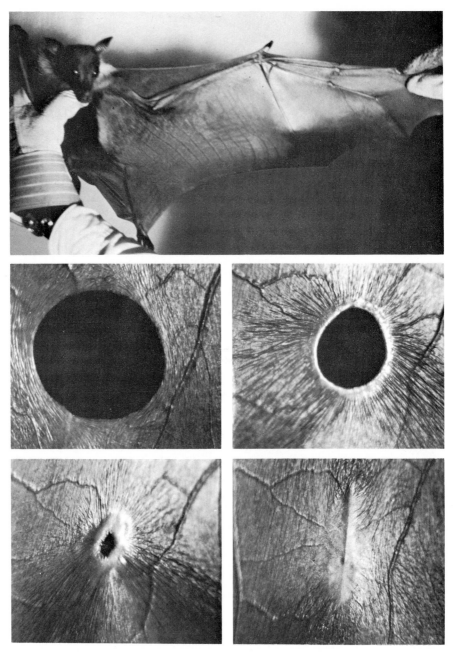

Fig. 104. When a 1.5-cm hole is punched through the wing membrane of the flying fox (*Pteropus*) (top), ingrowth of new tissues from the margins supplements contraction in obliterating the original aperture. Photographs of holes were taken 1 day, and 2, 3, and 5 weeks after injury.

webbing of ducks' feet, have been found not to elicit regenerative responses from the surrounding tissues. Therefore, regeneration of histologically complex parts of mammalian bodies is distributed in ways not easily explained.

In recent years surgeons have become aware that even man is not entirely without regenerative abilities. It has been shown that when the fingertips of children are accidentally lost, regeneration may follow provided that the injury is not covered by grafts of skin (Illingworth, 1974). When allowed to heal by epidermal migration, such fingers may produce outgrowths that even include the fingernail (Fig. 105). This type of regeneration diminishes in children more than 4 or 5 years of age and fails to occur at levels of amputation proximal to the last joint. Nevertheless, numerous examples of seemingly remarkable fingertip regeneration have been recorded, but whether this is a case of epimorphic regeneration or an exaggerated version of wound healing and tissue repair is for the future to decide. It may be worth noting, however, that the matrix from which the fingernail originates extends about 80% of the distance from the fingertip to the last joint, making it less probable that amputations would have removed all vestiges of the nail matrix (Goss, 1980). Reestablishment of the fingernail, therefore, may not

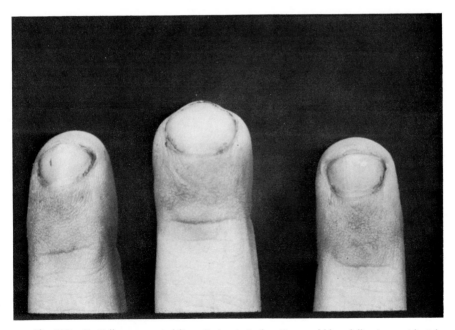

Fig. 105. Partially regenerated fingertip (center) of an 8-year-old boy following accidental amputation distal to the last joint 3 years earlier.

involve *de novo* regeneration. Inability to perform experiments on humans, or to study the histological sequence of events in purported fingertip regeneration, makes it unlikely that answers to crucial questions will soon be forthcoming, except in animal experiments. Borgens (1982), for example, has shown that infant mice, like human children, grow back the ends of their amputated digits.

Strategy and Prospects

From what fragmentary information is now available, it is possible to draw some conclusions about the factors that seem to have influenced regeneration loss during the course of vertebrate evolution, as well as its recovery in exceptional cases. First and foremost, the absence of regeneration seems to be correlated with the warm-blooded condition of birds and mammals. Homeotherms have a high metabolic rate that demands greater and more frequent food intake. Warm-blooded animals are therefore in danger of starving to death in a far shorter period of time than were their cold-blooded ancestors. It could be argued, therefore, that even if mammals were in fact able to regenerate legs, it would probably take several months to do so judging from the rates at which antlers are replaced. In this span of time, most herbivores would run a good chance of falling victim to predators, while the latter would probably starve if they were crippled. Thus, any mammals which might have acquired the capacity to regenerate legs would have benefited from little or no selective advantage. Instead, birds and mammals evolved a central nervous system efficient enough to enable them to avoid most losses before they occurred.

For many mammals, legs are vitally essential organs in the sense that the animal cannot long survive their loss the way many lower vertebrates can. The strategy of evolution would dictate that vitally essential organs ought not to be able to regenerate simply because the animal could not live long enough to replace them. It may be assumed, therefore, that if regenerative capacity were to evolve in birds and mammals it should be limited to structures that are not so important that their loss is fatal, nor so unimportant that they are not worth the effort to replace. The vulnerability of an organ might also have been a factor in the evolution of adaptive regenerative capacities. Clearly, antlers, ears, bat wing membranes, and children's fingertips are all vulnerable organs of nonvital importance. However, numerous other appendages in mammals could be similarly classified, but for unaccountable reasons they lack the ability to regenerate.

What are the prospects for inducing the regeneration of mammalian appendages not normally capable of replacing lost parts? Assuming there is a latent capacity for regeneration in the cells of all such structures, it would be necessary to remove or counteract experimentally whatever inhibiting factors may be blocking regrowth. The amputation stump should be healed over by epidermal migration, not sealed by wound contraction or grafts of full-thickness skin. This would encourage the intimate contact between epidermis and underlying mesodermal tissues upon which important inductive interactions are believed to depend, interactions that may be necessary for blastema formation, if not dedifferentiation of the cells from which such blastemas are derived.

Nonregenerating appendages form scars instead of blastemas. Latent regenerative ability might be enhanced if scar formation could be suppressed. This is not easily achieved in view of the ubiquity of collagen fibers and the efficiency with which scars seem to be produced in response to injuries. Nevertheless, the local administration of agents that inhibit collagen formation (e.g., cortisone) or of collagenase, an enzyme capable of digesting collagen, might be useful approaches to the problem.

Because nerves are so essential for the regeneration of appendages in lower vertebrates, they may also be important in mammals. The fact that nerves are not required for the regeneration of antlers (Wislocki and Singer, 1946), rabbit ear tissue (Grimes and Goss, 1970), or bat wing membranes (Church and Warren, 1968) does not mean that they might be unnecessary for the regeneration of more movable appendages, such as limbs or fingers. Therefore, a logical means of increasing the probability of regeneration would be to divert extra nerves from nearby locations into an appendage, hopefully to increase the nerve supply above the hypothetical threshold that might be required for regeneration. Attempts along these lines in young rats have not been successful (Bar-Maor and Gitlin, 1961). However, if regeneration of such hyperinnervated limbs were actually to occur in man, there would be serious problems of how the regenerate might be reinnervated. A limb's own nerves tend to grow back to appropriate end organs during the course of regeneration, ensuring reestablishment of normal sensation and motor activity. Foreign nerves would also be expected to regenerate, and would probably make synaptic connections with muscles in competition with the native innervation. Clinically, this could lead to chaotic motor coordination in which excess nerves might be worse than none at all. Solving the problem of mammalian regeneration will involve more than just promoting regrowth.

References

Bar-Maor, J. A., and Gitlin, G. G. (1961). Attempted induction of forelimb regeneration by augmentation of nerve supply in young rats. *Transplant. Bull.* **27,** 460–461.

Borgens, R. B. (1982). Mice regrow the tips of their foretoes. *Science* **217,** 747–750.

Church, J. C. T., and Warren, D. J. (1968). Wound healing in the web membrane of the fruit bat. *Br. J. Surg.* **55,** 26–31.

Goss, R. J. (1969). "Principles of Regeneration." Academic Press, New York.

Goss, R. J. (1978). "The Physiology of Growth." Academic Press, New York.

Goss, R. J. (1980). Prospects for regeneration in man. *Clin. Orthop. Relat. Res.* **151,** 270–282.

Goss, R. J. (1981). Tissue interactions in mammalian regeneration. *In* "Mechanisms of Growth Control" (R. O. Becker, ed.), pp. 12–26. Thomas, Springfield, Illinois.

Goss, R. J., and Grimes, L. N. (1972). Tissue interactions in the regeneration of rabbit ear holes. *Am. Zool.* **12,** 151–157.

Goss, R. J., and Grimes, L. N. (1975). Epidermal downgrowths in regenerating rabbit ear holes. *J. Morphol.* **146,** 533–542.

Grimes, L. N. (1974a). Selective x-irradiation of the cartilage at the regenerating margin of rabbit ear holes. *J. Exp. Zool.* **190,** 237–240.

Grimes, L. N. (1974b). The effect of supernumerary cartilaginous implants upon rabbit ear regeneration. *Am. J. Anat.* **141,** 447–451.

Grimes, L. N., and Goss, R. J. (1970). Regeneration of holes in rabbit ears. *Am. Zool.* **10,** 537.

Joseph, J., and Dyson, M. (1966). Tissue replacement in the rabbit's ear. *Br. J. Surg.* **53,** 372–380.

Illingworth, C. M. (1974). Trapped fingers and amputated finger tips in children. *J. Pediatr. Surg.* **9,** 853–858.

Kvinnsland, S., and Breistein, L. (1973). Regeneration of the cartilaginous nasal septum in the rat, after resection. Its influence on facial growth. *Plast. Reconstr. Surg.* **51,** 190–195.

McMinn, R. M. H. (1969). "Tissue Repair." Academic Press, New York.

McMinn, R. M. H. (1975). Electron microscopic observations on the repair of perforated tympanic membranes in the guinea-pig. *J. Anat.* **120,** 207–217.

Singer, M. (1951). Induction of regeneration of forelimb of the frog by augmentation of the nerve supply. *Proc. Soc. Exp. Biol. Med.* **76,** 413–416.

Singer, M. (1974). Neurotrophic control of limb regeneration in the newt. *Ann. N.Y. Acad. Sci.* **228,** 308–322.

Vorontsova, M. A., and Liosner, L. D. (1960). "Asexual Propagation and Regeneration" (Engl. Transl.). Pergamon, Oxford.

Wallace, H. (1981). "Vertebrate Limb Regeneration." Wiley, New York.

Wislocki, G. B., and Singer, M. (1946). The occurrence and function of nerves in the growing antlers of deer. *J. Comp. Neurol.* **85,** 1–19.

Abnormal Antlers

The magnificent morphologies represented by the antlers of many species of deer testify to the complexities of their mechanisms of development. However, the more complicated a process is, the more chances there are for mistakes to occur. Therefore, it is little wonder that the antlers of deer are so vulnerable to aberrations in size and shape. Yet such malformations as are known to occur shed light on how normal growth is controlled. Especially good illustrations of various types of antler abnormalities are to be found in Bubenik (1966) and Rörig (1901, 1907).

There are three possible causes of antler abnormalities (de Nahlik, 1959). One is of hereditary origin. The second is attributable to systemic conditions. A third group is caused by direct injury. It is not always obvious which is which.

Genetic Effects

Just as the normal size and shape of antlers is subject to genetic determination (Chapter 12), so also are certain irregularities. The problem is to determine which anomalies are genetic and which are not. Hereditary defects tend to occur in successive generations of deer. They are also reproduced each year as new sets of antlers are regenerated. They would be expected to occur bilaterally.

Peculiarities in the shapes of antlers are often the most distinctive features of individual deer. They are evidence of the genetic lability in a population. Indeed, the accumulation of such mutations has been responsible for the evolution of deer into the dozens of species that now exist. How much of antler morphology is genetically determined, and how much is of a more fortuitous nature, may be inferred from the differences that occur between the left and right antlers in an individual animal. No two antlers are exactly alike, although their similarities are often striking. Nevertheless, one antler sometimes possesses more tines than the other: its branches may be of different lengths, or they may sometimes grow out at variable angles. In the strategy of developmental genetics, only the more important things are coded in the genes. Unimportant details can be left to chance. Presumably these minor variations from one side to the other represent the results of unknown, but nongenetic, factors that influence the normal developmental processes by which antler growth is achieved. According to Baccus and Welch (1983), the incidence of antler asymmetry in sika deer increases with age.

In years past when deer parks flourished and gamekeepers knew their animals individually, aberrant antlers were recorded, not only from one year to the next, but from generation to generation. Such unusual antlers took many forms. Sometimes they were stunted, or missing altogether, as in hummels. In other cases they may have been crooked, or unbranched. Considered undesirable from the point of view of trophy hunting, such deer were often culled from the population, but their genes tended to be perpetuated covertly through the females in the herd.

In roe deer, the two antlers are on rare occasions observed to have coalesced at their bases (Fig. 106). In this species, the antlers grow vertically from the head, and their pedicles are normally closer together than in other types of deer. Whether or not the partial fusion of the two antlers represents the effects of a genetic predisposition is not proven, but Fooks (1955) described its occurrence in two successive years in one deer.

About 100 years ago, Caton (1884) described a white-tailed deer shot in Texas that carried palmate antlers on both sides (Fig. 107). Although one cannot rule out mechanical injury as the cause of this abnormality, its bilateral symmetry, coupled with the fact that another animal with similar antlers had been sighted in the same region, strongly suggests that this unusual configuration might have been the result of genetic mutation. The implications for the evolution of palmate antlers in moose, reindeer, and fallow deer are obvious.

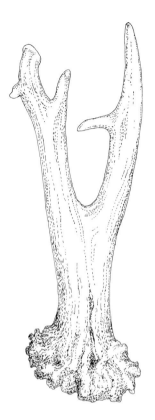

Fig. 106. When the pedicles of the roe deer are too closely approximated, their antlers may be partially fused. (After Nitsche, 1898.)

Sometimes deer have been known to grow extra antlers from ectopic sites. When these are not just accessory outgrowths from the normally located antlers, but occur in other parts of the body, it is difficult to attribute them to the effects of traumatic injury. In the white-tailed deer, for example, Nellis (1965) reported a supernumerary antler growing from the right zygomatic arch, although its association with the absence of the eye on the same side raises suspicions of injury-induced antlerogenesis, albeit from a part of the skull considerably removed from the normal region from which antlers are typically produced. In a mule deer observed over several years in Yosemite National Park, a third antler about 5 cm long sprouted medially from the nasal bones in front of the eyes. This extra antler was replaced in successive years, and a similar abnormality was reported in another deer in later years, suggesting the possibility of an inherited condition (Dixon, 1934). In the red

Fig. 107. Skull of a Texan white-tailed deer with abnormally flattened antlers. (After Caton, 1884.)

deer, Krieg (1956) described an unusual case of an animal with four antlers (Fig. 108). The extra pair protruded symmetrically below and anterior to the normal ones, and the similarities in their size and shape suggest the possibility of a genetic origin.

Systemic Influences

Certain abnormalities are sometimes observed to occur bilaterally, but not necessarily in successive years. The bilateral symmetry of such malformations argues against their being caused by mechanical trauma, and the fact that they are not necessarily repeated from year to year is consistent with the possibility of a nongenetic origin. Therefore, these are presumed to be the

Fig. 108. Example of bilateral duplicate antlers in a red deer, the accessory ones protruding from the skull below the normal antlers. (After Krieg, 1956.)

result of certain systemically distributed physiological influences, the nature of which is not always understood.

Curious malformations encountered in mule deer, red deer, and especially roe deer are corkscrew antlers (Krieg, 1954; Rhumbler, 1929; Rörig, 1907). This condition, which occurs bilaterally, is characterized by left-handed spirals on the left side and right-handed ones on the right (Fig. 109). It tends to be associated with lung worm infestations or tuberculosis, although the connection between pulmonary disease and the production of corkscrew antlers remains obscure.

Equally enigmatic are the stunted antlers grown by senile deer. Although remarkably similar antlers are grown each year by mature bucks, their last set of antlers, which forecasts the animal's impending demise, is typically abnormal. Such heads are said to "go back," and the condition is well established from observations of deer in captivity that have been allowed to live long enough to die of "old age." Whether the stunted antlers of elderly

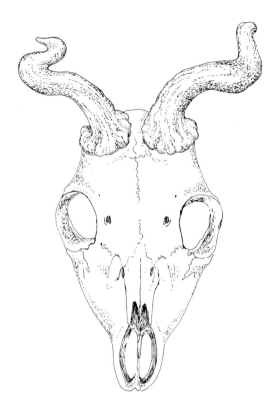

Fig. 109.   Corkscrew antlers in a roe deer. (After Rhumbler, 1929.)

deer are explained in terms of decreased sex hormone secretion, or are attributable to accumulated degenerative effects of old age, is not known. Such antlers bear no resemblance to the peruke antlers typically grown by castrated deer. They could be more closely associated with the stunted antlers sometimes produced under conditions of malnutrition (Chapter 12).

A different type of shortened antler is observed under experimental conditions in which the light regime has been artificially altered (Chapter 11). When the annual light cycle is shortened from 12 months to 3 or 4 months, for example, deer respond by growing antlers more frequently than normal. Subjected to abbreviated growth periods, these antlers mature precociously. Shortened outgrowths, sometimes amounting to little more than buttonlike antlers on top of the pedicles, are produced under these conditions (Goss, 1969a).

The normal regeneration of antlers each year depends in part on the temporal coordination between when the old antlers drop off and when the new ones make their debut. Although these two events are not causally related, the latter always follows or coincides with the former, presumably because each is triggered at the appropriate phase of the annual light cycle. However, when the photoperiod is atypical one might expect abnormalities in the antler cycle. Such is the case when deer are held under constant conditions of illumination. Lacking the usual fluctuations in day lengths that define the seasons of the year, deer tend to revert to their innate circannual rhythms of antler replacement (Chapter 11). In some animals subjected to these experimental conditions, new growth may start before the previous antlers have been cast, or vice versa. In the former case, the stump of the old antler may be pushed upward by its replacement (Fig. 110) before becoming detached belatedly from the growing tip. If this is prevented by screwing the uncast antler stump to the pedicle bone, new growth mushrooms out to the side, sometimes giving rise to eccentric branches (Fig. 111). In rare cases, the original antler may be cast, but renewed growth does not ensue. Even wound healing fails to occur, leaving an area of bare bone exposed on the end of the pedicle (Fig. 112).

A relatively infrequent abnormality involves uneven shedding of the vel-

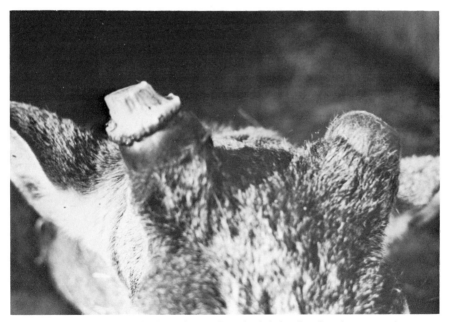

Fig. 110. Example of a sika deer whose right antler failed to detach as the new one began to grow. It was later dislodged.

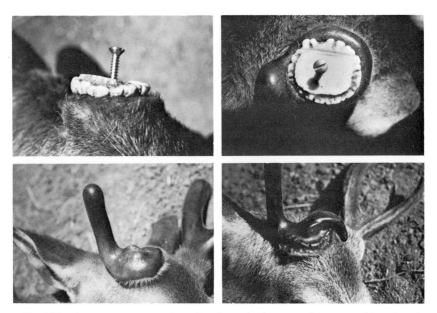

Fig. 111. Attempt to arrest casting of a sika antler by a tantalum screw driven into the pedicle bone on May 10. It did not prevent renewed growth, which bulged out on all sides (June 7) and sprouted a long branch posteriorly (June 28). When the screw and its surrounding bone were finally detached, an abortive outgrowth grew from the center of the pedicle (Sept. 6). (From Goss, 1963. Copyright 1963 by the American Association for the Advancement of Science.)

vet. It has been established that testosterone normally induces the skin to peel off following ossification of the antler at the end of the growing season. As would be expected, both antlers normally react to this hormonal influence simultaneously. When they do not, it is cause for bewilderment. Under unnatural lighting conditions, for example, a sika deer has been observed to cast one antler but not the other, or to grow a replacement only on one side (Fig. 113) (Goss, 1969b). Cases of red deer, illustrated by Bubenik (1966), show antlers which have shed their velvet distally while retaining it proximally. One can only explain such anomalies in terms of variable reactivities of local antler tissues to endocrine influences, or neural factors.

One of the most perplexing phenomena of antler growth is the purported effect on the development of the opposite antler following contralateral injury elsewhere on the body. Over the years there have been intermittent reports of deer injured in one way or another, usually in the legs, which subsequently developed abnormal antlers on one side of the body. Reports involving six different species of deer include some cases in which injuries

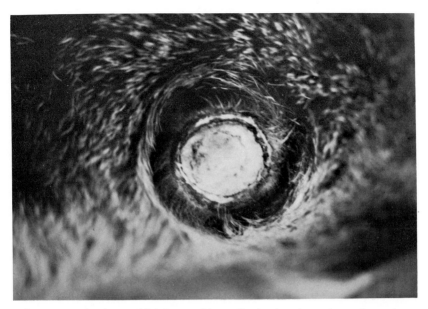

Fig. 112. Under abnormal lighting conditions, sika deer have been observed sometimes to lose their old antlers without growing new ones. In such cases the pedicle skin fails to heal over the exposed bone.

to one hind leg were correlated with antler abnormalities on the opposite side of the body (Acharjyo and Misra, 1972; Baillie-Grohman, 1894; Davis and Acharjyo, 1979; Marburger et al., 1972). When the injury was persistent, as in the case of leg amputation, the contralateral antler was shorter than normal for up to 6 successive years as illustrated in Figure 114 (by Davis and Acharjyo, 1979). However, in other cases, abnormal antlers have been observed on the same side as the injuries (Clarke, 1916; Moore, 1931). Gaskoin (1856) cited examples of fallow deer with unilateral antler abnormalities in the absence of disease or defects on either side of the body. In view of such inconsistent reports, it is prudent to remember Fowler's (1894) admonition that "fractures of the limbs are not uncommon in deer, and, apparently, abnormality of the antler is not uncommon; it is natural therefore that they should occasionally coincide in the same animal."

If true, it is difficult to explain the apparent contralateral effect except perhaps in terms of a compensatory response to counteract the imbalance caused by the original injury. A crippled deer would be expected to have an altered gate, the effects of which might result in compensatory changes in the growing antler owing to alterations in the flow or pressure of the blood supply. Morrison-Scott (1960) has suggested that as a deer turns to lick a wound, his outside antler would be more vulnerable to injury.

Fig. 113. Example of a sika deer, held under artificial equatorial (12L/12D) photoperiods, which cast and regenerated its left antler while retaining the bony one on the right side. (Goss, 1969b.)

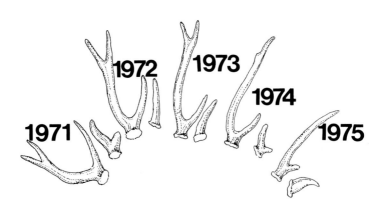

Fig. 114. Five successive pairs of antlers grown by a sambar deer between 1971 and 1975 following loss of its left hind leg in 1969. Each year the right antler was abnormally short, suggesting an effect of the contralateral injury. (After Davis and Acharjyo, 1979.)

The cases documented in the literature cannot be refuted. However, it is important to note that in the absence of controlled experiments, anecdotal evidence must be interpreted with caution. There are many cases in which uninjured deer grow unequal antlers, as well as instances in which normal antlers are produced despite injuries. There is sufficient evidence to suspect that the contralateral effect may be a valid phenomenon, but it deserves to be explored accurately and with adequate controls before it can be accepted as an established fact of antler growth.

Unilateral castration has also been claimed to affect the opposite antler (Darling, 1937; Matthews, 1952; Penrose, 1924). However, in this case, experiments have not confirmed the "cross effect." Penrose (1924) removed the right testis from a sika deer and noted that the left antler lacked a brow tine for the next 3 years. However, in another animal the left and right antlers were equal after hemicastration. Zawadowsky (1926) observed no influence on the antlers of red deer subjected to unilaterial castration or cryptochidism. In 1952, Jaczewski reported no effects of unilateral castration on the antlers of red deer and roe bucks. Thus, compelling evidence that loss of one testis can affect either antler seems not to have been established.

Injuries in Velvet

Like all developing systems, the growing antler is especially vulnerable to trauma. It is also endowed with capacities for regulation, because not all lesions result in permanent defects. The growing antler is richly supplied with sensory nerves, the principal function of which must be to enable the deer to avoid injury. The velvet hairs may serve primarily as tactile organs. The kinesthetic sense, whereby the locations of unseen parts of the body are monitored, is well developed. Even the largest species of deer can move rapidly through the obstacle courses of their native forests without injuring their tines. Deer feel uncomfortable when their velvet is touched, and typically pull away when one takes hold of their still growing antlers. Yet they have been observed to rub their growing antlers on their inguinal areas, or to "groom" them carefully with their hind hooves.

It is inevitable that accidents will happen. One of the most common injuries is the fractured antler (Nitsche, 1898; Pilkington, 1954). In extreme cases, the break is so severe that the distal part of the antler may hang pendulously to one side, attached only by the remaining velvet. If the blood supply is not disrupted, such broken ends might survive and continue to grow, albeit at a crooked angle (Fig. 115). In the case of nonunion of the fracture such segments are destined to be lost when the velvet is shed.

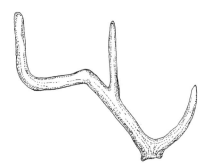

Fig. 115.   Fractured antler of red deer. Healing of the break occurred, resulting in a crooked antler, the end of which appears to have adjusted its angle of growth accordingly. (After Rhumbler, 1929.)

Occasionally an antler is only cracked, the distal portion maintaining its original orientation. This is typically followed by fracture healing, accompanied by a conspicuous swelling along the shaft in the region of the break. The histological details of this process have not been studied, but it would be interesting to compare the mechanisms by which antlers heal fractures with how other broken bones in the body repair themselves.

Sometimes the distal portion of an antler is lost altogether. This may be the result of a fractured antler in which the end portion does not survive, or the consequence of frostbite in the roe deer (which typically grows his antlers in winter). Growing antlers may be amputated surgically, as when velvet antlers are harvested for oriental markets (Chapter 15). If a tourniquet is not applied, the severed stump spurts blood from its arteries for considerable distances. However, the loss of blood is checked by the remarkable capacity of antler arteries to constrict at their severed ends, thus reducing hemorrhage to a trickle within minutes. Such amputated antlers readily heal their stumps. However, the extent to which they may regenerate has not been systematically explored. Nevertheless, it is well established that some regeneration is often possible (Goss, 1961; Tegner, 1954), but seldom if ever enough to replace the entire missing portions (Fig. 116). On deer farms, velvet antlers are sometimes harvested early to allow second growths to occur in the same season. Presumably the regenerative capacity of developing antlers is affected by the level of amputation and the stage of antler development at which the operation is performed. It is probably also a function of the species of deer. In the reindeer, for example, the amputation of a tine is occasionally followed by the regeneration of a new one. Indeed, mechanical trauma to the shaft of the reindeer antler sometimes induces the outgrowth of a branch at the injured location (Bubenik, 1956). The re-

Fig. 116. Amputation of the growing tip of an antler is sometimes followed by regeneration of a short point. (From Goss, 1961.)

generative ability of growing antlers is apparently made possible by the persistence of antlerogenic tissue for varying distances proximal to the growing tip along the main beam. Like the cambium of plants, this periosteal layer of tissue (Fig. 87) may to varying degrees possess the latent capacity for morphogenetic development.

The growing tip of an antler is soft enough to be easily incised. Bubenik and Pavlansky (1959) cleaved antler buds sagittally or transversely and

Fig. 117. Double antler growth from a single pedicle can be promoted by resecting a wedge of tissue (May 6) from the antler bud of a sika deer (a), each half of which gives rise (June 3) to its own branch(es) (b). (From Goss, 1961.)

noted how readily the two halves grew together again, even when they attempted to keep them separated with methylmethacrylate wedges (which tended to be extruded). However, if a sufficient gap is maintained between the two halves of an antler bud (Fig. 117), double antlers may develop (Bubenik and Pavlansky, 1959; Goss, 1961). Such injuries may explain the occasional natural occurrence of double antlers bifurcated above the base, as reported in moose (Caton, 1877) and other species (Merk-Buchberg, 1916). Thus, as in embryonic limb buds and regeneration blastemas, considerable morphogenetic regulation is possible following various experimental interventions. That is, whole structures can develop from portions of a bud, as they can from fused ones, in accordance with the expression of morphogenetic fields.

Pedicle Wounds

As in other regenerating systems capable of expressing autotomy, the maximal potential for growth resides immediately proximal to the breakage plane. While distal levels of the antler have limited capacities for regeneration, only the pedicle is capable of giving rise to a complete and normal antler. Although the antler develops from the distal end of the pedicle stump, experiments have shown that if the pedicle is previously amputated it is still capable of giving rise to an antler (Goss, 1961). It can also do so in reversed polarity from a grafted segment of pedicle (Goss, 1964). Resection of the entire pedicle, along with a generous amount of the surrounding frontal bone, is not sufficient to abolish altogether the antlerogenic potential (Bubenik et al., 1956). It is clear that the antler territory extends well beyond the portion of the pedicle that normally gives rise to the antler. Antlers produced following partial or total removal of the pedicle are usually abnormal and often lack brow tines, but they may also become overly large (Bubenik et al., 1956). Nevertheless, the persistence of antlerogenic potential in regions of the pedicle or nearby cranium raises the interesting question of why the latent capacity of these tissues to give rise to antlers is normally held in abeyance.

Numerous cases have been reported of accessory antlers produced from the pedicle or on nearby regions of the skull (Bird, 1933; Hartwig, 1969; Nitsche, 1898; Rörig, 1901; Tilak, 1978; Whitehead, 1955). In some of these cases the onset of antler outgrowth was triggered by an injury to the pedicle or adjacent frontal bone. Such lesions can apparently initiate antler growth, which is consistent with the established fact that regeneration cannot occur except in association with a healing epidermal wound. Yet lesions to the pedicle do not always induce accessory antler growth. Usually the wound heals and nothing grows out. What subtleties are involved in determining whether pedicle trauma will or will not give rise to an accessory antler remain to be determined.

Typically, accessory antlers are abnormally developed. They are invariably shorter than normal, sometimes being little more than abbreviated spikes. Nevertheless, they tend to follow the usual antler replacement cycle, the velvet being shed in the fall and the antlers dropping off in the spring when regrowth begins.

Injuries to the pedicle skin may have interesting effects on the early stages of regeneration. The missing skin of a circumcised pedicle is readily regenerated but the resulting antler bud may be conspicuously swollen (Fig. 118) because of subdermal accumulation of fluid (Bubenik et al., 1956; Goss, 1961). Presumably this effect is the result of interference with the circulatory pattern in the pedicle, such that venous drainage is temporarily occluded. It is a transient phenomenon, but one which emphasizes the importance of the circulation in normal antler development.

Fig. 118. Edematous antler bud produced by a sika deer following excision of its pedicle skin prior to the onset of growth. This effect, presumed to result from vascular disturbances, was corrected in subsequent growth. (From Goss, 1964. Reprinted with permission from Pergamon Press, Ltd.)

Do Nerves Affect Antler Development?

In view of the rich innervation of growing antlers, and the importance of nerves in the regeneration of appendages by lower vertebrates, it is important to explore the possibility that nerves might exert influences on antler growth. In 1946, Wislocki and Singer succeeded in denervating the antlers of white-tailed deer. Although this operation did not prevent further growth of the antlers, the final products were stunted and abnormal in shape. This effect was attributed to the desensitization of the denervated antlers such that the deer could not avoid bumping them on objects, thus resulting in the observed abnormalities. Although denervation experiments have not subsequently been repeated, experiments have been carried out more recently on the effects of electric stimulation on antler growth. When constant electric current was applied to growing mule deer antlers, the results were shortened and misshapened antlers (Lake et al., 1978, 1983). However, when Bubenik et al. (1982) applied alternating current to the antler nerves, there was a 70% increase in elongation and a 40% increment in the weight of the antlers subsequently produced. As in the earlier experiment, such antlers were abnormal in shape and were not necessarily enlarged in subsequent years.

An even more intriguing effect of nerves on antler growth has been proposed by Bubenik et al. (1982) as a result of their observations of the effects of trauma to the growing antlers of white-tailed deer. When one of the animals accidentally split his young growing antler, not only did the injured antler respond by becoming exaggerated in size, but the uninjured opposite one reacted in similar fashion, but to a lesser degree (Fig. 119). The giant antlers grown by this deer were reproduced in successive years (Fig. 120). This intriguing effect could conceivably be caused by the persistence of injured tissues in the pedicle from which subsequent antlers were grown, although this could not explain the contralateral effect. Alternatively, Bubenik et al. (1982) proposed that the excessive antler growth might have been induced by the innervation of the antlers such that the original trauma promoted a neurotrophic influence not only responsible for the overgrowth on the injured side but for the vicarious response on the opposite uninjured antler. The repetition of antler overgrowth in subsequent years was attributed to a "memory" effect mediated through the central nervous system. This possibility was tested by performing similar injuries on anesthetized deer. Under these circumstances, in which the deer were not acutely subjected to the pain of injury, no abnormalities in subsequently developing antlers were observed. Therefore, it was concluded that the pain experienced in connection with the injury to the growing antler may so condition the brain as to promote antler overgrowth not only on both sides of the head but also in future years.

The role of trauma in antler induction has also been disclosed by experiments designed to induce antler growth in female deer (Chapter 14). Jaczewski (1977) demonstrated that if pedicles were caused to form in female red deer by hormonal manipulation, antler development from these pedicles could be stimulated by amputating their tips. Without such amputation, no

Fig. 119. Accidental injury to the 10-cm right antler of a white-tailed deer in velvet was followed by exuberant growth on both sides, more so on the right. (Courtesy of Dr. George Bubenik.)

Fig. 120. Same animal as in Figure 119, but 3 years later. The antlers have grown abnormally large each time they were replaced. (Courtesy of Dr. George Bubenik.)

antlers developed. Similarly, in female wapiti it has been shown that trauma, either surgical or chemical, to the presumptive antler pedicle region, can trigger the production of antler growth from the affected sites (Robbins and Koger, 1981). Bubenik et al. (1982) demonstrated that in white-tailed does pretreated with antiandrogen, surgical injury to the skin and underlying periosteum would stimulate antler growth. Thus, female deer clearly possess the potential for antler production, a potential that can sometimes be activated by local injury. Such injuries probably exert their influences directly on the antlerogenic periosteum coupled with the induction of wound healing in the overlying epidermis.

Nonexistent Mistakes

What does not occur in nature is sometimes as significant as what does. Although growing antlers are vulnerable to misdirected development, it is instructive to be aware of such abnormalities as are rarely if ever observed. For example, the overall size of an antler is not only species specific, but is evidently related to the size of the deer, or to that of the pedicle from which the antler is produced. Although the antlers of an individual animal may be small at young ages and larger in mature animals, one does not encounter diminutive or enlarged antlers without the corresponding morphology having been affected. In other words, antlers of fully mature shape, with the typical number of points, are not known to occur in miniature, nor in extra large dimensions. If the size of the antler is reduced, the number of points is correspondingly decreased, and vice versa.

Another feature not subject to abnormality, at least under natural conditions, relates to the orientation of antlers. Antlers are either right or left, and their axes of orientation are always commensurate with those of the skull from which they grow. One does not find antlers growing backward or sideways in nature.

Although the distal regions of a growing antler may sometimes be missing as a result of the premature cessation of elongation, the opposite situation does not occur. That is, distal portions do not develop in the absence of the more proximal regions. This is in keeping with the normally proximodistal order in which antlers develop.

Typically, antlers require 3–5 months to grow. Except for Père David's deer, this applies to all species, from the smallest to the largest. Bigger antlers grow faster, completing their development in the prescribed period of time. Even in castrated deer, whose antlers remain permanently viable and in velvet (Chapter 13), growth ceases and resumes at the usual times of year. Thus, constantly growing antlers have neither been observed in nature nor produced experimentally, probably for the same reasons that unlimited growth of other organs and appendages does not take place. It is conceivable that mutations for such hypothetical abnormalities, having no survival value, did not persist in the gene pool. Indeed, the overgrowth of those appendages that develop by terminal addition (as opposed to internal expansion) would of necessity require the production of new parts on their ends. In the absence of genetic information required for the development of such supernumerary components, it would be difficult to imagine how these abnormalities could occur in the first place. In some types of deer, such as *Odocoileus*, there seems to be no predetermined number of points that their antlers are capable of forming. In others (e.g., sika deer), the maximum number of points per antler almost never exceeds four. These comparisons emphasize how little is understood about the inheritance of antler morphology. Perhaps there are no genes directly determining the number of points per antler. If not, the diameter of the main beam may be limited by body size, with the overall length of the antler and its number of points a derivative of this.

Another condition thus far not definitively diagnosed in antlers is cancer. Although the velocity at which antlers elongate far exceeds the rates most cancers grow, antlers are remarkably free from neoplasia. In view of their transient lifespans, this apparent absence of cancer in antlers may simply reflect the fact that most tumors require more than a few months to develop. Clearly, the relation between antler growth and tumor development deserves to be explored, especially in the perennial antlers grown by castrated deer. Although such antlers tend to develop amorphous tuberosities after their first year if prevented from freezing in the winter, the nature of these unlimited growths, and their possible relationship to neoplastic development, is an aspect of antler biology still awaiting exploitation.

Abnormalities need not be morphological. They may also affect the seasonality of the antler cycle. Except when subjected to artificially altered light cycles or to hormonal imbalance, deer in temperate zones are remarkably synchronized in their rhythms of antler loss, regrowth, and maturation. Deer that normally grow antlers in the summer, for example, are not observed to do so in winter. Even in those species that typically grow winter antlers (e.g., roe deer, Père David's deer), what may have originated as an abnormality became the norm when everyone else did it.

Horn Abnormalities

Horns, like antlers, are subject to injury and defect. Not being renewable structures, however, their deformities are persistent. As described in Chapter 3, horns grow by the deposition of successive conical increments of keratin inside those previously formed. The parts produced earliest in life are pushed farther and farther outward as new material is laid down proximally. In all but one species of horned ruminants, these cornified segments build up over the years making it possible to determine the age of an animal, at least in temperate zones, by counting the annual rings. However, in the pronghorn antelope the horn sheath is shed annually and replaced by a new outgrowth each spring. If the pronghorn buck is castrated, he is unable to shed the horn sheaths each year, and successive annual increments of horn are built up which curl forward in an abnormal simulation of what typically occurs in all other horned animals (Fig. 41).

Unlike antlers, horns differentiate from the integument. Although they secondarily involve the underlying cranial bone which develops into the horn core, it is in the skin that the horn originates in the newborn animal. If this skin is removed, cauterized, or destroyed by the application of caustic chemicals, horn development is prevented or seriously stunted. In polled animals, the absence of horns is hereditary (Hammond, 1950).

Various lengths of a horn may sometimes be broken off, in which case the missing part is not replaced. Occasionally a horn may be broken at the base, but remain attached and viable. This results in a permanently askew appendage. Wild animals maintained in zoos are particularly vulnerable to such breakages when their untamed temperaments collide with the unyielding conditions of their confinement.

Horns sometimes reduplicate themselves. This may occur naturally in such animals as the 4-horned sheep (Fig. 121), but it is an abnormal condition when two horns are produced from the same region of the head, presumably resulting from the subdivision of the horn-forming tissues. Ectopic horns sometimes occur in other regions of the body, but usually on the head. These may be sizable structures, or small scurs of cornified epidermis. They are usually loose and unattached to the underlying skull. Such patho-

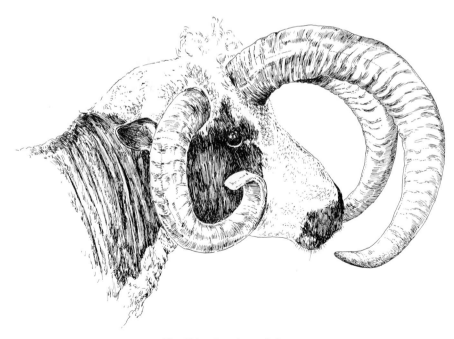

Fig. 121. Four-horned sheep.

Fig. 122. Early illustration of a horned man. (After Avalon, 1958.)

Fig. 123. Capture of a unicorn, duped into submission by a virgin. (After Davis, 1976.)

logical growths are sometimes encountered even in animals not normally bearing horns. They have been reported growing from the ear of a horse (Miller, 1917), from the head of a man (Fig. 122) (Avalon, 1958), and from the forehead of a woman (Avalon, 1958; Wood, 1917). Whether such freaks are homologous with the horns normally occurring in ruminants is a matter for conjecture.

Unicorns exist only in myths and legends (Davis, 1976; Dove, 1936; Hathaway, 1980; Ley, 1948). Real horns are normally grown by artiodactyls, or even-toed ungulates. However, most renditions of unicorns show the horn growing from the forehead of a horse, a perissodactyl. Yet ancient artists, in an unconscious concession to zoological accuracy (Fig. 123), have often provided their otherwise equine subjects with cloven hooves! Whether or not they grew horns in both sexes seems not to have been of concern. Unification of paired horns has been artificially achieved in goats and cattle by surgical fusion of the presumptive horn skin at the midline of the forehead shortly after birth. This results in the production of a single horn (Fig. 42) formed by the fusion of the two that would otherwise have grown, thereby simulating the mythical unicorn (Dove, 1935, 1936).

Horns are susceptible to cancer. Neoplasms are known to occur, particularly around the base of a horn, which may result in considerable disfigurement. Squamous cell carcinomas are not uncommon in association with horns, and have been known to metastasize to the lungs (Kulkarni, 1953; Pachauri and Pathak, 1969). Deer have unwittingly escaped such pathological risks by evolving annual instead of perennial appendages on their heads.

References

Acharjyo, L. N., and Misra, R. (1972). Effect of amputation of a hind limb on the growth of antlers of deer in captivity. *Indian For.* **98,** 507–508.

Avalon, J. (1958). La corne dans la médecine, l'art et la littérature. *Aesculape* **41,** 3–57.

Baccus, J. T., and Welch, R. D. (1983). Asymmetry in the antler structure of sika deer (Cervus nippon) from the Edwards plateau of Texas. *In* "Antler Development in Cervidae" (R. D. Brown, ed.). Caesar Kleberg Wildl. Res. Inst., Kingsville, Texas (in press).

Baillie-Grohman, W. A. (1894). Abnormal red-deer heads. *Field* **2150,** 356.

Bird, R. D. (1933). A three-horned wapiti (*Cervus canadensis canadensis*). *J. Mammal.* **14,** 164–166.

Bubenik, A. B. (1956). Eine seltsame Geweihentwicklung beim Ren. *Z. Jagdwiss* 2(1), 21–24.

Bubenik, A. B. (1966). "Das Geweih." Parey, Hamburg.

Bubenik, A., and Pavlansky, R. (1959). Von welchem Gewebe geht der eigentliche Reiz zur Geweihentwicklung aus? III. Mitteilung: Operative Eingriffe am Bastgeweih. *Saugetierkd. Mitt.* **7,** 157–163.

Bubenik, A.,Pavlansky, R., and Rerabek, J.(1956). Trophik der Geweihbildung. Z. Jagdwiss. **2,** 136–141.

Bubenik, G. A., Bubenik, A. B., Stevens, E. D., and Binnington, A. G. (1982). The effect of neurogenic stimulation on the development and growth of bony tissues. J. Exp. Zool. **219,** 205–216.

Caton, J. D. (1877). "The Antelope and Deer of America." Hurd & Houghton, Cambridge, Massachusetts.

Caton, J. D. (1884). Abnormal deer antlers from Texas. Am. Nat. **18,** 733–737.

Clarke, F. C. (1916). Malformed antlers of deer. Calif. Fish Game **2,** 119–123.

Darling, F. F. (1937). "A Herd of Red Deer. A Study of Animal Behaviour." Oxford Univ. Press, London/New York.

Davis, T. A. (1976). Alicorn or the horn of the unicorn. Sci. Rep. April, pp. 227–230.

Davis, T. A., and Acharjyo, L. N. (1979). Effect of limb amputation on antler size in sambar deer. Pr. Mater. Zootech. **19,** 101–107.

de Nahlik, A. J. (1959). "Wild Deer." Faber & Faber, Ltd., London.

Dixon, J. S. (1934). A study of the life history and food habits of mule deer in California. Part I. Life history. Calif. Fish Game **20,** 181–282.

Dove, W. F. (1935). The physiology of horn growth. A study of the morphogenesis, the interaction of tissues, and the evolutionary processes of a Mendelian recessive character by means of transplantation of tissues. J. Exp. Zool. **69,** 347–406.

Dove, W. F. (1936). Artificial production of the fabulous unicorn. A modern interpretation of an ancient myth. Sci. Mon. **42,** 431–436.

Fooks, H. A. (1955). Coalesced roe heads. Field **205,** 536.

Fowler, G. H. (1894). Notes on some specimens of antlers of the fallow deer, showing continuous variation, and the effects of total or partial castration. Proc. Zool. Soc. London pp. 485–494.

Gaskoin, J. S. (1856). On some defects in the growth of the antlers, and some results of castration in the Cervidae. Proc. Zool. Soc. London **24,** 151–159.

Goss, R. J. (1961). Experimental investigations of morphogenesis in the growing antler. J. Embryol. Exp. Morphol. **9,** 342–354.

Goss, R. J. (1963). The deciduous nature of deer antlers. In "Mechanisms of Hard Tissue Destruction" (R. F. Sognnaes, ed.), Publ. No. 75, pp. 339–369. Am. Assoc. Adv. Sci., Washington, D.C.

Goss, R. J. (1964). The role of skin in antler regeneration. Adv. Biol. Skin **5,** 194–207.

Goss, R. J. (1969a). Photoperiodic control of antler cycles in deer. I. Phase shaft and frequency changes. J. Exp. Zool. **170,** 311–324.

Goss, R. J. (1969b). Photoperiodic control of antler cycles in deer. II. Alterations in amplitude. J. Exp. Zool. **171,** 223–234.

Hammond, J. (1950). Polled cattle. Endeavour **9,** 85–90.

Hartwig, H. (1969). Versuch zur Analyse der Entwicklungsbedingungen, die zu einer Doppelstrangen-Bildung beim Reh fuhrten. Z. Jagdwiss. **15,** 167–169.

Hathaway, N. (1980). "The Unicorn." Viking Press, New York.

Jaczewski, Z. (1952). Effect of unilateral castration on antlers growth in roe-deer (Capreolus capreolus L.) and common stag (Cervus elaphus L.). Fragm. Faun. Mus. Zool. Pol. **6,** 199–205.

Jaczewski, Z. (1977). The artificial induction of antler cycles in female red deer. Deer **4,** 83–85.

Krieg, H. (1954). Rehkronen. Wild Hund **56,** 408–410.

Krieg, H. (1956). Ein Hirsch mit doppelseitiger Geweih-Missbildung. Z. Jagdwiss. **2,** 217–220.

Kulkarni, H. V. (1953). Carcinoma of the horn in bovines of the old Baroda State. Indian Vet. J. **29,** 415–421.

Lake, F. T., Davis, R. W., and Solomon, G. C. (1978). The effects of continuous direct current on the growth of the antler. *Am. J. Anat.* **153,** 625–630.

Lake, F. T., Davis, R. W., and Solomon, G. C. (1983). Bioelectric phenomena associated with the developing deer antler. *In* "Antler Development in Cervidae" (R. D. Brown, ed.). Caesar Kleberg Wildl. Res. Inst., Kingsville, Texas (in press).

Ley, W. (1948). The legend of the unicorn. *In* "The Lungfish, the Dodo, and the Unicorn," pp. 19–34. Viking Press, New York.

Marburger, R. G., Robinson, R. M., Thomas, J. W., Andregg, M. J., and Clark, K. A. (1972). Antler malformation produced by leg injury in white-tailed deer. *J. Wildl. Dis.* **8,** 311–315.

Matthews, L. H. (1952). "British Mammals," pp. 280–304. Collins, London.

Merk-Buchberg, M. (1916). Mehrstangigkeit und Geweihbildung bei weiblichen Cerviden. *Zool. Beobachter.* **57,** 98–102.

Miller, J. E. (1917). Horned horses. *J. Hered.* **8,** 303–305.

Moore, W. H. (1931). Notes on antler growth of Cervidae. *J. Mammal.* **12,** 169–170.

Morrison-Scott, T. C. S. (1960). Antler anomalies. *J. Mammal.* **41,** 412.

Nellis, C. H. (1965). Antler from right zygomatic arch of white-tailed deer. *J. Mammal.* **46,** 108–109.

Nitsche, H. (1898). "Studien über Hirsche." Engelmann, Leipzig.

Pachauri, S. P., and Pathak, R. C. (1969). Bovine horn cancer: Therapeutic experiments with autogenous vaccine. *Am. J. Vet. Res.* **30,** 475–477.

Penrose, C. B. (1924). Removal of the testicle in a sika deer followed by deformity of the antler on the opposite side. *J. Mammal.* **5,** 116–118.

Pilkington, R. (1954). Damaged antlers. *Country Life* **115,** 523.

Rhumbler, L. (1929). Zur Entwicklungsmechanik von Korkziehergeweihbildungen und verwandter Erscheinungen. *Wilhelm Roux Arch. Entwicklungs mech. Org.* **119,** 441–515.

Robbins, C. T., and Koger, L. M. (1981). Prevention and stimulation of antler growth by injections of calcium chloride. *J. Wildl. Manage.* **45,** 733–737.

Rörig, A. (1901). Über Geweihentwickelung und Geweihbildung. IV. Abnorme Geweihbildungen und ihre Ursachen. *Arch Entwicklungs Mech. Org.* **11,** 225–309.

Rörig, A. (1907). Gestaltende Correlationen zwischen abnormer Körperkonstitution der Cerviden und Geweihbildung derselben. *Arch. Entwicklungs Mech. Org.* **23,** 1–150.

Tegner, H. (1954). Malformed antlers. *Country Life* **116,** 577.

Tilak, R. (1978). A case of a third antler in barking deer (Cervidae; Mammalia). *Deer* **4,** 265.

Whitehead, G. K. (1955). Deformities of antlers. *Country Life* **117,** 991.

Wislocki, G. B., and Singer, M. (1946). The occurrence and function of nerves in the growing antlers of deer. *J. Comp. Neurol.* **85,** 1–19.

Wood, R. H. (1917). A woman with horns. *J. Hered.* **8,** 434.

Zawadowsky, M. M. (1926). Bilateral and unilateral castration in *Cervus dama* and *Cervus elaphus. Trans. Lab. Exp. Biol. Zoo Park Moscow* **1,** 18–43.

The Case of the
Asymmetric Antlers

Left and right antlers are normally mirror images of each other. The re-markable similarity between the sets of antlers produced by the same adult buck year after year is equally impressive (Fig. 138). Therefore, when the two antlers of a deer are different from one another, it is cause for wonder. Such a situation occurs in reindeer and caribou antlers. Their brow tines, which grow down over the snout, are usually developed as vertically flat-tened branches on one side or the other, the opposite one remaining as an unbranched spike (Fig. 124). This asymmetry of the brow tines involves enlargement of the left side more often than the right. Sometimes both left and right brow tines are branched or palmate, and sometimes neither is. One may logically ask what function the enlarged brow tines serve, why this enlargement predominates on the left side, and whether or not the asymme-try is genetically determined.

The Prevalence of Asymmetry

Brow tines are a special case of a more general phenomenon in which the usual bilateral symmetry of organisms is out of balance. Indeed, the pre-

Fig. 124. Caribou with nearly mature antlers illustrating the palmate left brow tine and the unbranched right one.

dominance of right-handedness in man, perhaps the best known example of asymmetry, is still more speculated about than understood.

Other examples of biological asymmetry abound (Ludwig, 1932; Neville, 1976; Oppenheimer, 1974). Whenever a growing vine climbs around a pole, it must spiral in either a clockwise or counterclockwise direction (Davis, 1974b). The whorls of petals in many flowers overlap each other in one direction or the other. The claws of lobsters and fiddler crabs are unilaterally enlarged, sometimes the asymmetry being distributed at random, other species being all left-handed or all right-handed. Male narwhals grow a long straight tusk that develops from the left cannine tooth, and also spirals to the left. However, in rare instances, specimens have been found that possess two tusks by development of both canines; but, in such cases, each tusk still spirals to the left (Fig. 45).

More familiar cases of asymmetry are encountered in our own anatomy (Corballis and Morgan, 1978; Morgan and Corballis, 1978). The heart is displaced to the left in the chest cavity, presumably reflecting its own postnatal left-sided enlargement to accommodate the functional inequities of the two sides of this organ. Related to cardiac asymmetry are the unequal sizes of the two lungs whereby the left one is smaller, and is composed of fewer lobes, than the right. In snakes this laterality is carried to the extreme. Many species possess only one lung because of the failure of the left lung to develop beyond rudimentary proportions. This would appear to be an adaptation to the elongate body plan of such animals. It is not uncommon for one gonad to be larger than the other, particularly in the case of the ovaries. The most extreme example of this is the fact that birds possess only one ovary, which is on the left side. The organ that might otherwise have developed into the right ovary remains rudimentary from early embryonic stages. Indeed, removal of the left ovary results in the belated development of the right gonad, but in such cases it becomes a testis instead of an ovary, reversing the animal's sex with its asymmetry (van Limborgh, 1970). In many bats, only one ovary is functional, perhaps as insurance against the possibility of producing twins, a potentially detrimental attribute in animals adapted for flight.

Asymmetries arise when it is disadvantageous, if not impossible, for an organism to preserve a basic balance between its left and right sides. Right-handedness, for example, is a solution to the inconvenience of having to decide which hand to use if they were equal. This is reminiscent of the arbitrary convention of putting the North Pole at the top of the map when there is clearly no absolute up or down in the universe. In a more biological vein, it is noteworthy that all proteins are composed of amino acids of the levorotatory configuration, instead of the equally plausible dextrorotary alternative. Because life on earth could not have it both ways, a random event in the early stages of evolution became an irreversible commitment. If there is life elsewhere in the universe, and if it happens to utilize the same amino acids that we use but of the opposite mirror image, these two forms of organisms could not survive by eating each other because their molecules would be sterically incompatible.

Is Asymmetry Genetic?

The expression of asymmetry can be explained by three mechanisms. First, it may be a result of random choices, in which case left and right inequities would be expected on average to occur in equal numbers. Second, it could be genetically determined, as is presumably the case when most of all members of a population are asymmetric in the same direction. Finally, the development of left–right asymmetry may be influenced by environmental factors which are themselves not in balance. Reindeer and caribou antlers provide a unique opportunity to obtain basic information about the possible genetic control of asymmetry.

If asymmetries are inherited, it is difficult to understand the developmental mechanism by which this might occur. Genes operate by controlling the synthesis of molecules, which in turn become incorporated into biological systems. Because it is inconceivable to consider genes themselves as being left- or right-handed, one must ask how they can be responsible for bilateral asymmetry. However, it is equally difficult to comprehend the genetic basis of bilateral symmetry itself except insofar as even the unfertilized egg is endowed with its own bilateral organization. Be this as it may, we can only assume that unknown physiological and developmental mechanisms exist to tip the scales one way or the other.

The ideal system in which to analyze the genetic basis of asymmetry would be one in which the unequal parts could be induced to develop more than once in the same organism. Under such circumstances, reversal of the original asymmetry would be compelling evidence against the strictly genetic control of left or right inequalities. Several opportunities for this approach are provided by certain invertebrate organisms, especially crustaceans. As mentioned earlier, lobsters and fiddler crabs possess chelae, one of which is larger than the other. In all such instances, they are of equal size during larval stages, but acquire their inequities during subsequent molts. These appendages also have the ability to regenerate following amputation. However, when induced to grow back, such claws faithfully reproduce themselves, thus preserving the original asymmetry. However, there are some crustaceans in which asymmetry is reversible (Cheung, 1976; Mellon and Stephens, 1978; Wilson, 1903). In the stone crab and pistol shrimp, for example, removal of the larger of the two chelae results in the transformation of the smaller (unamputated) claw into the larger type at the next molt. The missing larger one is replaced with a smaller regenerate.

This phenomenon, known as compensatory regulation (Zeleny, 1905), is also encountered in a small marine annelid worm (*Hydroides*) which dwells in calcareous tubules secreted around themselves on the surfaces of rocks and shells. These worms can plop back into their tubes when disturbed, plugging the opening with an umbrellalike operculum. This operculum is actually a modified gill filament, and is functional and fully developed only on one side at a time. One-half the population have functional right opercula, the other half have functional left ones. The contralateral operculum remains rudimentary. However, if the functional one is cut off, instead of regenerating itself in place, the opposite rudimentary one then develops into a functional operculum (Schochet, 1973a,b). This reversal of asymmetry is apparently a normal part of the animal's existence, and can lead to repeated reversals as often as functional opercula may be lost. The mechanisms responsible for the reversals of asymmetry in annelids and crustaceans have defied all attempts by experimental zoologists to explain this intriguing phenomenon.

The Brow Tine Mystery

The brow tines of reindeer and caribou (Fig. 124) provide an unparalleled opportunity to analyze the potential genetic basis of this asymmetry (Goss, 1980). Because they regenerate a new set of antlers each year, it is possible to compare successive sets of brow tines in the same individual. The basic question is whether or not antler asymmetry is preserved from year to year. First, it is worth asking what the function of the brow tine might be, and what is the relative frequency with which the left and right brow tines are enlarged.

The function of the brow tines remains to be explained. They may serve the same function as do the brow tines of other deer, namely, as offensive weapons that also protect the head in combat (Bubenik, 1975). However, this does not seem consistent with the fact that females as well as males exhibit their enlargement even though females do not fight the way males do. It has also been suggested that the palmate configuration of the brow tine may be an adaptation for shoveling snow in order to reach lichens beneath. Antlers have never been observed to be used in this fashion, and even if they were, the brow tine would be in an extremely awkward position to function as a snow shovel. Besides, males cast their antlers in early winter (Davis, 1973). Reindeer and caribou use their hooves, not their antlers, to dig out craters in the snow (Murie, 1935). Another possibility is that the flattened brow tines might serve to deflect branches from the eyes when woodland caribou thresh their heads back and forth in the underbrush (Pruitt, 1966). However, even on the tundra, caribou still possess enlarged brow tines (Freeman, 1968). Alternatively, they may serve as a visual signal to other animals by enlarging the apparent size of the head as seen in profile. Finally, one wonders if they interfere with stereoscopic vision, and if so what the advantage of this could be.

Unhappily, none of the aforementioned notions explains the true purpose of these interesting structures. Future biologists, challenged to speculate on the subject, would do well to take into account the possible relationship between enlarged and asymmetric brow tines to some of the other unique attributes of reindeer and caribou. These include the fact that they live farther north than any other deer, form larger and better organized herds, migrate greater distances, are plagued by more flies, and grow antlers in both sexes. It may also be worth questioning the relationship between the palmate shape of the brow tine and the occurrence of this configuration only on one side, usually the left. Finally, any attempt to explain the function of brow tines should take into account the pronounced tendency for them to grow inward toward the midline of the snout.

There are four basic types of reindeer and caribou antlers classified according to their brow tines. These may be designated by using upper case and lower case letters to indicate branched (or palmate) brow tines, and unbranched (digitate) or absent ones, respectively. Thus, an animal with an enlarged left brow tine and a small or absent right one is referred to as Lr. The opposite situation is lR. Symmetric antlers are indicated by LR or lr, depending on whether both brow tines are enlarged or both are small or absent.

The distribution of enlarged brow tines is not even (Table II). Tabulations of 380 pairs of reindeer and caribou antlers reported in various publications reveals that in 200 cases (52.6%) the left brow tine was enlarged and the right was not (Lr). The reverse situation (lR) was encountered in 113 cases (29.7%). Therefore, over 80% of the deer carried asymmetric antlers. Bilateral enlargement of both brow tines (LR) occurred in 57 cases (15.0%), and only 10 (2.6%) had neither brow tine enlarged (lr). However, it may be worth noting that the last figure may be spuriously small. In view of the fact that these data are based mostly on specimens in zoos and museums where there is obvious selection in favor of older and larger animals carrying more impressive antlers, there is reason to suspect that younger animals tend to have smaller brow tines. Nevertheless, the fact remains that significantly more reindeer and caribou are left dominant (Lr) than right dominant (lR).

It is possible that what happens on one side might affect development of the opposite brow tine. However, the fact that the incidence of bilaterally enlarged brow tines (15.0%) approximates so closely the product of the left dominant and right dominant ones (52.6% Lr $\times$ 29.7% lR = 15.6% LR) strongly suggests that the two brow tines develop quite independently of one another. There is simply a higher incidence of brow tine enlargement on the left side than on the right (67.6% versus 44.7%). Again, the product of these percentages yields a 30.2% incidence of enlarged brow tines per individual, equivalent to a predicted 15.1% of paired antlers with both brow tines

TABLE II

Distribution of Brow Tine Asymmetry in Reindeer and Caribou Inhabiting the Northern Hemisphere[a]

Types of antlers	Banfield (1954)	Davis (1973, 1974a)	Goss (1980)	Murie (1935)	Totals
Lr	24 (68.6%)	123 (58.3%)	46 (42.2%)	7 (28.0%)	200 (52.6%)
lR	8 (22.9%)	66 (31.3%)	32 (29.4%)	7 (28.0%)	113 (29.7%)
LR	3 (8.6%)	22 (10.4%)	21 (19.3%)	11 (44.0%)	57 (15.0%)
lr	0	0	10 (9.2%)	0	10 (2.6%)

[a] As documented by various researchers (Goss, 1980).

enlarged. This figure is remarkably close to the 15.0% of LR cases observed. Therefore, there would seem to be neither inhibitory nor stimulatory effects exerted by one brow tine on the other; each grows independently large or small in accordance with whatever factors are responsible for controlling the shape and size of these tines.

We know neither the functions of brow tines, nor why their asymmetries favor the left side. Yet is has been possible to learn whether or not successive sets of antlers produced by the same animal reproduce their original asymmetry (Bergerud, 1976; Goss, 1980). The answer has come from a longitudinal study of individual reindeer and caribou available in captivity. In some cases only 2-year sequences have been obtained, but in one case the antlers were followed for 7 consecutive years. In all, 10 animals have been studied in various locations from Alaska to Boston, for an average of 3.8 years per deer. In many cases the ages of the deer were not known, and all too often the longitudinal study was terminated by the death of the deer. Nevertheless, 28 2-year sequences have been recorded for 5 males and 5 females. They confirm the fact that these deer are capable of switching their asymmetries from year to year (Table III).

The antler types remained the same in 11 of these cases (39.3%), but on the other 17 occasions (60.7%) the antler types were altered from one year to the next (Fig. 125). In most of the latter cases, only one of the two brow tines changed its type, but on four occasions both of them were altered simultaneously. In general, there was a tendency for deer with two small brow tines to develop large ones. There was also a tendency for the animals to become increasingly dominant on the right side. With advancing age,

TABLE III

Successive Types of Brow Tines Grown by Ten Reindeer and Caribou Observed for 2–7 Years Each[a]

	Males					Females				
Years	1	2	3	4	5	6	7	8	9	10
1	Lr	Lr	LR	Lr	lr	lr	lr	Lr	lr	Lr
2	LR	IR	LR	Lr	IR	IR	IR	Lr	lr	lr
3	LR		Lr	Lr	IR	IR	Lr	Lr	lr	
4	IR				LR			IR	Lr	
5	LR				IR				LR	
6	LR								IR	
7									Lr	

[a] Goss, 1980.

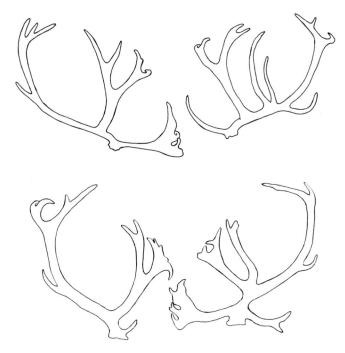

Fig. 125. Reversal of brow tine asymmetry in successive sets of antlers from the same reindeer. (After Goss, 1980.)

digitate brow tines tend to be replaced by enlarged ones, and left dominant sets of antlers become dominant on the right side. Therefore, the predisposition of reindeer and caribou to be left dominant may reflect the prevalence of this condition in younger animals, animals which under normal circumstances are more numerous in a wild population.

Although brow tine morphology does not necessarily remain the same from one year to the next, genetic control cannot be completely ruled out. The occurrence of small tines in younger animals, for example, may be the result of genetic influences. The apparent shift of asymmetry from left to right with increasing age may also have a genetic basis. Nevertheless, the fact remains that reindeer and caribou exhibit a strong tendency to be left dominant, at least in early life. However, inheritance cannot account for the unpredictability with which antler asymmetry is expressed. Therefore, one must consider environmental factors which might influence the laterality of brow tine development. It may be significant that all of the reindeer and caribou thus far considered have been residents of the northern hemisphere. However, there exist herds of reindeer on South Georgia Island (50°–55° S)

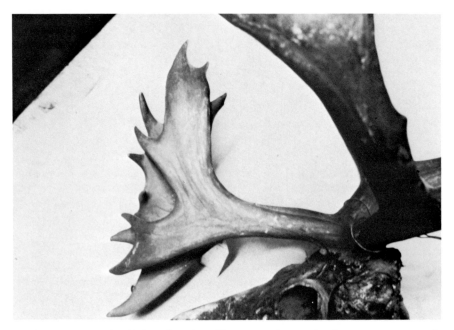

Fig. 126. Example of bilaterally enlarged brow tines showing the alternate arrangement of their points, none of which coincides with one from the opposite side.

which were introduced there from Norway early in this century as a potential source of meat for whaling ships (Leader-Williams, 1978). Tabulations of 70 specimens of these southern hemisphere reindeer have revealed that 27.1% were left dominant and 38.6% right dominant (N. Leader-Williams, personal communication). Therefore, it is possible that antler asymmetry may somehow be affected by Coriollis forces created by the rotation of the earth. Indeed, there is evidence of such influences in the case of coconut palms which spiral predominantly to the right in southern latitudes and to the left north of the equator (Davis, 1974b).

It remains to be determined how the size and shape of reindeer and caribou brow tines are controlled. Such influences as nerves or blood flow suggest themselves. Although some of the foregoing data indicate that the two tines develop independently, there is also some intriguing evidence that their development may be coordinated. For example, in those instances in which both brow tines are palmate, they grow out parallel to each other but in close proximity. When the points are formed on their ends, they are more often than not found to interdigitate (Bubenik, 1975), giving the appearance

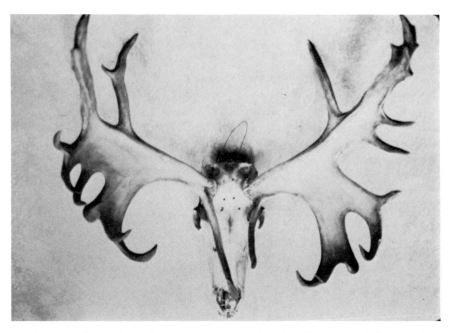

Fig. 127. The reindeer's dominant brow tine on the right side angles medially across the snout, while the digitate left brow tine grows straight forward. This could be interpreted as a means of shifting the center of gravity of the antlers closer to the midline.

of a pair of folded hands (Fig. 126). It is a curious thing that the developing tines do not "bump" into each other. In other cases, in which the brow tine is enlarged on only one side, it exhibits a strong tendency to grow toward the midline of the face (Fig. 127). At the same time, the smaller unbranched tine grows straight forward, almost as if it were keeping its distance from the larger one. Indeed, when there are two enlarged brow tines, there is a lesser tendency to bend medially. One cannot escape the impression that these outgrowths are somehow reacting to each other's presence in adjusting their directions of growth. Nowhere is this phenomenon more convincingly illustrated than in certain rare instances in which both brow tines are palmate, but one angles dorsally, the other ventrally, each bending medially so as to line up one above the other in a vertical plane (Allen, 1908). Thus, the two brow tines each grow into what would otherwise be one-half a tine, coordinating their morphologies so as to give the impression of a single large structure (Fig. 128). It is difficult to speculate about what mechanisms must be operating to achieve such a curious configuration.

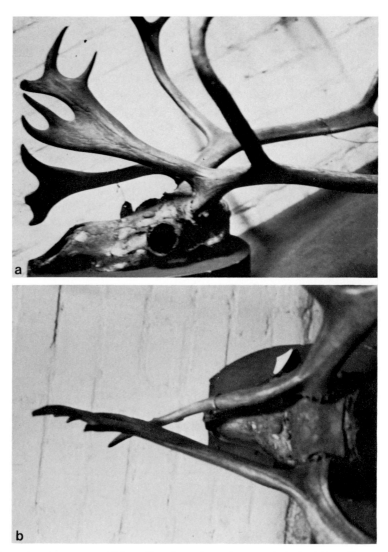

Fig. 128. A rare case of bilaterally palmate brow tines in which the left one has grown above the right and vice versa (a), the two becoming aligned medially with each other (b).

References

Allen, J. A. (1908). The Peary caribou (*Rangifer pearyi* Allen). *Bull. Am. Mus. Nat. Hist.* **24,** 487–504.

Banfield, A. W. F. (1954). Preliminary investigation of the barren ground caribou. Part II. Life history, ecology, and utilization. *Wildl. Manage. Bull., Ser. 1* **10B,** 1–112.

Bergerud, A. T. (1976). The annual antler cycle in Newfoundland caribou. *Can. Field Nat.* **90,** 449–463.

Bubenik, A. B. (1975). Taxonomic value of antlers in genus *Rangifer,* H. Smith. *Proc. Int. Reindeer/Caribou Symp., 1st, 1972* pp. 41–63.

Cheung, T. S. (1976). A biostatistical study of the functional consistency in the reversed claws of the adult male stone crabs, *Menippe mercenaria* (Say). *Crustaceana* **31,** 137–144.

Corballis, M. C., and Morgan, M. J. (1978). On the biological basis of human laterality. I. Evidence for a maturational left-right gradient. *Behav. Brain Sci.* **2,** 261–269, 277–336.

Davis, T. A. (1973). Asymmetry of reindeer antlers. *Forma Functio* **6,** 373–382.

Davis, T. A. (1974a). Further notes on asymmetry of reindeer antlers. *Forma Functio* **7,** 55–58.

Davis, T. A. (1974b). Enantiomorphic structures in plants. *Proc. Indian Natl. Sci. Acad., Part B* **40,** 424–429.

Freeman, M. M. R. (1968). Ethnozoological interpretation of the brow-tine in Arctic caribou. *Arct. Circ.* **18**(3), 45–46.

Goss, R. J. (1980). Is antler asymmetry in reindeer and caribou genetically determined? *Proc. Int. Reindeer/Caribou Symp., 2nd, 1979* pp. 364–372.

Leader-Williams, N. (1978). The history of the introduced reindeer of South Georgia. *Deer* **4,** 256–261.

Ludwig, W. (1932). "Das Rechts-Links-Problem im Tierreich und Beim Menschen." Springer-Verlag, Berlin/New York.

Mellon, D., Jr., and Stephens, P. J. (1978). Limb morphology and function are transformed by contralateral nerve section in snapping shrimps. *Nature (London)* **272,** 246–248.

Morgan, M. J., and Corballis, M. C. (1978). On the biological basis of human laterality. II. The mechanisms of inheritance. *Behav. Brain Sci.* **2,** 270–336.

Murie, O. J. (1935). "Alaska-Yukon Caribou," North Am. Fauna No. 54. U.S. Dept. Agric., Washington, D.C.

Neville, A. C. (1976). "Animal Asymmetry." Arnold, London.

Oppenheimer, J. M. (1974). Asymmetry revisited. *Am. Zool.* **14,** 867–869.

Pruitt, W. O., Jr. (1966). The function of the brow tine in caribou antlers. *Arctic* **19,** 111–113.

Schochet, J. (1973a). Opercular regulation in the polychaete *Hydroides dianthus* (Verrill, 1873). I. Opercular ontogeny, distribution and flux. *Biol. Bull. (Woods Hole, Mass.)* **144,** 400–420.

Schochet, J. (1973b). Opercular regulation in the polychaete *Hydroides dianthus* (Verrill, 1873). II. Control of opercular regulation. *J. Exp. Zool.* **184,** 259–280.

van Limborgh, J. (1970). The primary asymmetry of the gonadal primordia in the duck. *Z. Anat. Entwicklungs Gesch.* **130,** 37–79.

Wilson, E. B. (1903). Notes on the reversal of asymmetry in the regeneration of the chelae in Alpheus heterochelis. *Biol. Bull. (Woods Hole, Mass.)* **4,** 197–210.

Zeleny, C. (1905). Compensatory regulation. *J. Exp. Zool.* **2,** 1–102.

Light and Latitude

Deer are adaptable creatures. They inhabit all latitudes from the equator to the arctic, and are found on four continents. The most southern species is the huemul, an inhabitant of the Andes in southern Chile (ca. 50° S). At the other extreme are the reindeer and caribou adapted for survival on the tundra. Peary's caribou is an all white native of Ellesmere Island at 79° N latitude. Norwegian reindeer have been introduced to Svalbard where they survive at equally high latitudes. In the far north it is absolutely essential that reproduction be precisely timed to optimal seasons, because natural selection ruthlessly eliminates those that do not conform. Equatorial deer are not concerned about summer versus winter. Although the rainy and dry seasons may affect the survival of fawns, most species of tropical deer can breed all year around. However, in the temperate zones it becomes important for deer to be able to time their mating season to ensure that their young are born in the spring. Like most other animals, deer cannot afford to wait until the weather changes to begin to adapt to new seasons. By then it is too late. Instead, they lock onto the annual changes in day length as the most accurate environmental indicator of the changing seasons.

In 1925, William Rowan proved that light is more important than temperature in the seasonal adaptations of animals. He prevented juncos from migrating south by holding them in outdoor cages in Manitoba over the winter, supplementing the normally short days with extra illumination. Despite the extremely low temperatures during the winter of 1924, these birds developed enlarged gonads at the wrong time of year in response to the artificially lengthened days. Since this breakthrough, numerous other birds and mammals have been shown to depend on the annual light cycle as the most reliable environmental cue to the time of year. This applies not only to the reproductive cycle, but also includes such diverse phenomena as molting, seasonal color changes, behavior, food consumption, migration, and hibernation. In the case of deer, the annual replacement of antlers is particularly sensitive to the seasonal changes in day length, as is the reproductive cycle with which it is coordinated.

This was originally discovered by Jaczewski (1954) in Poland in the early 1950s. Working with the red deer, he locked some animals in a dark shed from 4:00 PM to 8:00 AM every day, beginning in the spring when the deer had just started to grow new antlers. They were therefore abruptly switched from the increasing day lengths characteristic of the Polish spring to 8-hour days such as prevail in the winter at that latitude. This had a dramatic effect on the growth of the antlers. Their elongation ceased, they became heavily mineralized, and the velvet was shed, all changes characteristic of what normally happens in the late summer in preparation for the autumn breeding season. Yet these events occurred in the spring, and by June or July the experimental deer had become so unmanageable that it was no longer possible to herd them into the shed each day. Accordingly, they were again exposed to natural outdoor lighting conditions which were approximately twice as long as the artificially shortened days to which the deer had recently been exposed. As a result, the animals reacted as if spring had come again. They cast the first sets of antlers they had just grown and regenerated a second set in what remained of the summer. These results showed for the first time that deer are not only capable of growing more than one set of antlers per year, but that they do so in response to the fluctuations in the photoperiod.

Some years later, the pioneering investigations of Jaczewski were followed up with a series of long-range studies in America on how the Japanese sika deer responds to a variety of artificial photoperiods (Goss, 1983). In these investigations, several methods of altering the annual light cycle were explored. The phase was shifted, the amplitude altered, and the frequency increased or decreased. It has even been possible to shorten or lengthen the normal 24-hour day.

Reversed Seasons

The simplest intervention is to reverse the light cycle, thus simulating a shift between the northern and southern hemispheres. What happens to animals transported across the equator is confirmed by their responses to the experimental reversal of the annual light cycle (Goss, 1969a). They change their reproductive periods and antler replacement cycles to conform to the altered environment. Over the years quite a few deer species have been imported from Europe, Asia, and America to New Zealand (Donne, 1924). They have also been established in certain locations in Australia. Reindeer have been introduced to South Georgia Island (Leader-Williams, 1978) in the South Atlantic and to Kerguelen in the Indian Ocean. Such deer have successfully adapted to the reversed seasons of the southern hemisphere.

Most deer typically grow their new antlers in the spring when the day lengths are increasing. They shed the velvet when the days are decreasing in late summer. Although it would seem obvious that it might be the lengthening of the day that initiates casting and regrowth of new antlers, this has not been confirmed experimentally. Indeed, the older males of some species lose their antlers before the winter solstice.

In an attempt to determine more precisely how the antler growth cycle might be affected by increasing versus decreasing day lengths, two reciprocal experiments were conducted on sika deer by Goss (1976). One group was held on progressively decreasing day lengths for a period of 3 years, the other on increasing photoperiods for the same length of time (Fig. 129). The first group was started on long days, 20 hours of light (L) and 4 hours of dark (D), but at 4-month intervals thereafter the days were decreased and the nights increased by 2 hours. At the end of 3 years, the light regime had become 4L/20D. The other group of deer was started on 4L/20D and was exposed in stepwise fashion to 2-hour increments in the light period every 4 months until they reached 20L/4D. The 4-month interval between successive changes in the light schedules was chosen so as not to coincide with the 6-month alternation between increasing and decreasing photoperiods in the outdoor environment, but as a period of time long enough to allow antlers to develop between successive changes in illumination. The basic reaction to these conditions was that the deer replaced their antlers at 8-month intervals, correlated with every other time the photoperiods were changed. This occurred regardless of whether the day lengths were successively increased or decreased. However, it was noted that in those animals subjected to stepwise decreases in photoperiod, the replacement of their antlers temporarily ceased to occur when the light schedule reached 12L/12D, but resumed as the light and dark periods again became unequal. Conversely, deer on successive increases in photoperiod remained longer in velvet, and

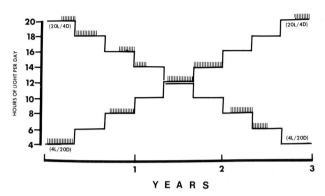

Fig. 129. Stepwise increases and decreases in the photoperiod at 4-month intervals caused antlers of sika deer to be replaced at 8-month intervals. However, at the equinox the deer skipped a cycle on decreasing light, but remained longer in velvet on increasing photoperiods. (After Goss, 1976.)

replaced their antlers more frequently during that part of the experiment when the days and nights became equal. Like the other group, they reverted to antler replacement cycles every other time the photoperiods were changed during the latter part of the experiment when the light to dark ratio became greater than one.

The fact that deer tend to replace their antlers every other time the length of day changes, irrespective of the direction of that change, suggests that the normal casting and regrowth of antlers in the spring may not be attributed directly to the increasing day lengths of that time of year. If deer normally grow a new set of antlers every other time the day length changes, which is once a year, the usual replacement of antlers in the spring may simply coincide with the fact that this is also the time of year when the fawns are born. To test this hypothesis, one could contrive to have fawns born in the autumn instead of spring, and note whether or not they would subsequently replace their antlers in the fall or spring (Goss, 1980). This hypothesis was put to the test by introducing pregnant female sika deer to artificially re-versed light cycles in the spring. As a result, their fawns were born into a light cycle in which the day lengths were decreasing instead of increasing, i.e., 6 months out of phase from the outdoor environment. After weaning, their mothers were returned outdoors, but the fawns were kept on the ar-tificially reversed light cycles. As expected, they grew their first sets of antlers at approximately 1 year of age when the days were again decreasing in length. This supports the fact that the first set of antlers is programmed to develop in yearlings independent of the prevailing light cycles (Chapter 6). However, the second sets of antlers were grown during the artificial spring

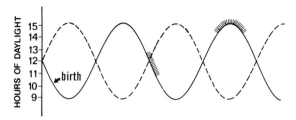

Fig. 130. When sika deer were born into an artificial autumn (—), they grew their first antlers in the next "autumn" at about 1 year of age. These antlers were cast in the following "spring" at the approximate age of 1.5 years, when the days were lengthening. Outdoor light cycle (---). (After Goss, 1980.)

when the animals were approximately 1.5 years old, rather than in the artificial autumn when they would have been 2 years old (Fig. 130). These results disprove the original hypothesis that the onset of antler growth is not affected by the direction of photoperiodic changes. Casting and regrowth of antlers would seem to be triggered directly by increasing day lengths after all, regardless of when birth occurs. This makes ecologic sense because sometimes fawns are born in late summer or early fall following winter conceptions. Yet these deer, if they survive, eventually synchronize with the rest of the herd, thus avoiding maladapted reproductive cycles.

Frequency Changes

The planet Earth happens to have a 365-day year. Because its axis is tilted about 23° with respect to the plane of orbit around the sun, the sidereal year is punctuated with seasons at nontropical latitudes. In adapting reproductive cycles to these seasons, animals native to temperate zones have evolved a responsiveness to the annual fluctuations in day lengths. The question arises as to whether they are reacting to the passage of a year per se or to the light cycle by which a year is normally defined. To test these alternatives, it is necessary to expose animals to light cycles with periods unequal to 365 days (Fig. 131).

As Jaczewski (1954) originally demonstrated, deer will produce two sets of antlers when exposed to two light cycles per year. This has been confirmed in sika deer held on light cycles of 6 months duration (Goss, 1969a), which amounts to doubling the frequency with which the annual changes in the photoperiod normally occur. As expected, deer grew two sets of antlers under these conditions in each 12-month period. In order to learn how far one can push this reaction, other groups of sika deer have been exposed to a

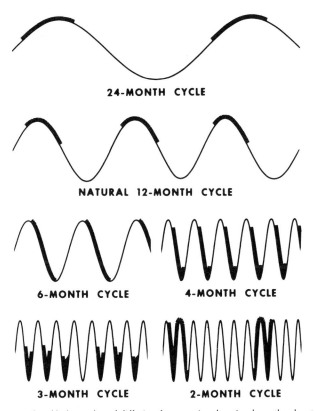

24-MONTH CYCLE

NATURAL 12-MONTH CYCLE

6-MONTH CYCLE   **4-MONTH CYCLE**

3-MONTH CYCLE   **2-MONTH CYCLE**

Fig. 131. Family of light cycles of differing frequencies showing how the duration in velvet is both shortened and shifted to the right with decreasing periods. (From Goss et al., 1974.)

threefold acceleration of the annual light cycle, that is, conditions in which the period of a cycle equals 4 months. These deer grew three sets of antlers per year. When maintained on four cycles per year, the limit of the deer's adaptability was approached. They underwent antler replacement cycles every 3 months three times in a row, but skipped the fourth time around before replacing their antlers on schedule in the fifth cycle. Finally, when the annual periodicity was reduced to only 2 months, deer failed altogether to respond to this sixfold acceleration of the light cycle. Instead, they waited until the normal time in the spring to cast and regrow their antlers.

When deer are maintained under the above unearthly conditions, they clearly demonstrate their dependence on the alternating increases and decreases in the annual light cycle as the primary environmental factor coordinating their reproductive cycles with the seasons of the year. It normally takes up to 4 months to grow a set of antlers. Therefore, approximately one-third of the year is spent in velvet, and during the other two-thirds the antlers are composed of dead bone. These characteristics are correlated with the infertile condition of the deer during the spring and summer when they are in velvet, and the fertile phase in the fall and winter. When the annual rhythm of changing day lengths is accelerated to the point where a cycle is equal to or shorter than the length of time normally required to grow a set of antlers, some kind of accommodation must be effected. Unless the rate at which antlers elongate were increased in proportion to the shortened period in velvet, stunted antlers would be grown. Because deer are unable to change the growth rates of their antlers, the shorter the light cycle the shorter the antlers. On a 6-month cycle, the antlers remain in velvet little more than one-half the normal growth phase and grow to about 40% of normal lengths. When the cycle is cut to 4 months, the antlers may attain lengths of only 5–15 cm, or 25% of normal, before the changing photoperiod causes their premature hardening, followed by shedding of the velvet. Under these conditions, the deer spend about 2 months of the 4-month cycle in velvet, and the other 2 months with polished antlers. On a 3-month cycle, antlers may grow 3–11 cm (Fig. 132) in the first or second episode, but their

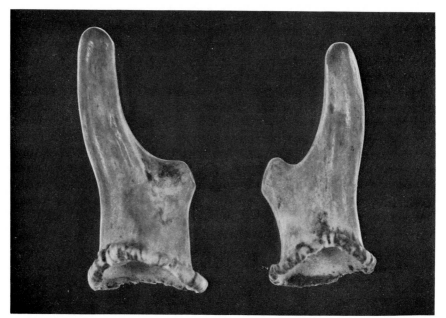

Fig. 132.   Abbreviated antlers grown by a sika deer held under a 3-month "annual" light cycle. (From Goss, 1969a.)

elongation is aborted soon after growth commences in later cycles. Consequently, these antlers are little more than flattened "buttons" on top of the pedicles. On such a 3-month "year" about 1 month is spent in velvet, and 2 months with bony antlers (Fig. 133).

This seems to be the limit of the deer's adaptive capabilities. The reason is to be found in the timing of antler growth with respect to the phase of the light cycle. In nature, the onset of antler growth usually occurs between the vernal equinox and the summer solstice when the days are still lengthening. The cessation of antler growth, and shedding of the velvet, occurs as the days are growing shorter at the end of the summer, but usually before the autumn equinox. When the duration of a cycle is shortened, casting and regrowth of antlers does not occur until shortly after the artificial summer solstice. The growing antlers may therefore remain in velvet throughout the phase of shortening days. On cycles of 4 or 3 months, the velvet phase of the antler replacement cycle shifts even farther to the right, until it occurs almost at the opposite times of year. This may explain why deer cannot grow more than three or four sets of antlers a year: one cycle does not have time to complete itself before the next begins. At the same time, the total duration of the velvet phase of antler growth is compressed into a period of about 1 month

In order to discover how the antler cycle might react to extra long years, a group of sika deer was maintained on a 24-month light cycle. Their reactions were found to vary according to the age of the deer. Those animals that had been introduced to this prolonged light cycle as yearlings replaced their

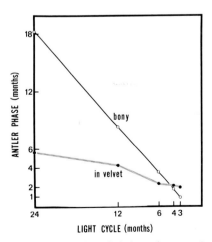

Fig. 133. Effects of various lengths of yearly light cycles on periods when sika deer antlers are bony (—○—) or in velvet (···●···). (After Goss, 1969a.)

antlers on a 24-month cycle. Although they remained in velvet for up to 6 months, the antlers they produced were not extra large. They reached their full size at the scheduled time, but remained in velvet without further elongation for about 2 months more. In contrast to yearlings, adult deer exposed to a 24-month light cycle replaced their antlers every 12 months during the 2-year duration of the experiment. These results suggest that the prior conditioning of animals to natural environmental conditions may affect their responses to artificial light cycles. An innate biological rhythm may ensure the annual replacement of antlers despite certain extreme lighting conditions. When the period of the light cycle is too short (2 months) or too long (24 months), deer revert to a cycle of about 12 months.

The Role of Latitude

The amplitude of the annual day length curve varies from 24 hours of light (or dark) per day, as occurs above the Arctic or Antarctic Circles, to the complete absence of seasonal changes in the lengths of day and night on the equator where there is a year-round equinox. Between these two extremes, the lengths of days and nights at the solstices increase with latitude.

There has been one published study of the reproductive seasons of deer as a function of latitude. This was conducted by Fletcher (1974), who collated the calving seasons of red deer in both hemispheres from data gathered in various zoos located between 41°–57° N and 26°–38° S. This investigation showed that not only was there no difference in the average times of birth at these ranges, but there was also no difference in the lengths of the calving seasons at the latitudes included in the study. There may be two explanations for these findings. One is that the range of the study may not have extended close enough to the equator. Another is that zoo animals are usually displaced from their native habitats, with no guarantee that such animals would adapt to a new latitude the way endemic species have.

The effects that distance from the equator has on reproductive and antler cycles of different species of deer in the wild are not easily defined, if only because so many types of deer do not inhabit a sufficiently extensive range of latitudes. However, it is fortunate that the very deer exhibiting the widest north–south habitat is the one for which the most information is available. Scattered data from a wide variety of sources on members of the genus *Odocoileus* have been plotted at appropriate latitudes on the map of North America in Figure 134. Viewed in this way, it is possible to compare the degree of variability prevailing at different latitudes from Manitoba (Ransom, 1966) to Mexico (Leopold, 1959; Mearns, 1907). Included are records of when white-tailed and mule deer are in rut, cast their antlers, give birth to

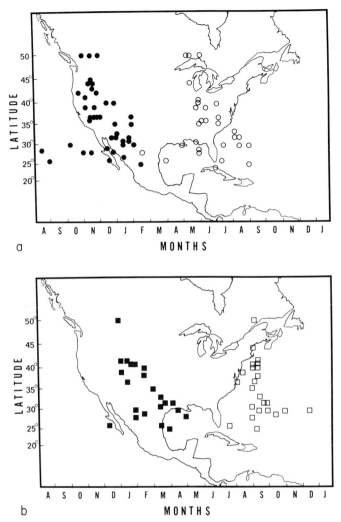

Fig. 134. Scattergram of average dates when white-tailed and mule deer are in rut (●) or give birth (○) (a), and cast their antlers (■) or shed their velvet (□) (b) as a function of latitude. Map is superimposed only as a reminder of latitude, not to indicate longitude. Data were assembled from several dozen separate reports on the genus *Odocoileus*.

fawns, and shed their velvet. Despite the variations that plague such analyses, certain interesting trends emerge.

It will be noted that in the northern extent of their range, these deer breed and give birth, and cast and regrow their antlers, in close synchrony with the seasons. Occasional individuals may stray from the pattern, but they are quickly "weeded out" of the population by a hostile environment. Below latitudes in the low thirties, approximately across the southern tier of states from California to Georgia, the reproductive and antler events become noticeably more dispersed in time. The seasons of mating and birth are less rigidly confined by adverse weather conditions, and there is more flexibility in when the antlers are cast and the velvet shed. The latter events seem not to exhibit so much spread as rutting and parturition, perhaps because they are directly controlled by the photoperiod, whereas successful breeding depends more on behavior and chance encounters of males and females. If a doe fails to become pregnant the first time she is in heat, she can wait a few weeks until the next estrus, or the one after that. The fawning dates are correspondingly scattered from spring into summer. There seems to be a slight drift to the right at progressively lower latitudes, especially in the antler cycles.

Although the dates of the equinox and solstice are the same everywhere, the magnitudes to which the lengths of days and nights are stretched at the summer and winter solstices are markedly increased at higher latitudes. This tends to define more sharply the times when various events in the year of the deer take place. Indeed, under artificial lighting conditions it has been noted that when the day lengths are changed abruptly rather than gradually, experimental deer react in a more synchronized fashion with respect to the casting and regrowth of their antlers (Goss, 1969a). The more gradual changes in day length at lower latitudes may therefore elicit responses spread over a longer period of time. Carried to its probable extreme, this could extend the seasons of rut, birth, and antler growth throughout the entire year as tropical latitudes are approached. This has occurred in Guatemala (15° N) (Handley, 1950). It would appear that the transition from the seasonal mode of reproduction (and antler replacement) of the temperate zone to the aseasonal one of the tropics may be a gradual one.

The Tropical Paradox

The various seasonal changes that occur in deer, including the replacement of their antlers, are entrained to the alternating increases and decreases in day length during the course of the year. The question of how deer might react to year-round constant day lengths is therefore an interesting one, the more so for the existence of deer whose native habitat is on or near the equator. Tropical deer differ from their counterparts in temperate zones by being able to breed at any time of year. At higher latitudes, in contrast, there have been selective pressures to concentrate breeding into specific seasons that are more narrowly defined the farther away from the equator the deer live. Presumably it was the absence of such selective pressures that enabled the temperate zone ancestors of today's tropical species to expand their breeding seasons throughout the calendar. Indeed, if the reproduction of such deer were not limited to specific times of year, the species would not be restricted to the production of one birth per year. Tropical deer should be able to give birth more frequently than once a year because there is nothing to prevent them from becoming pregnant again as soon as they have weaned their previous fawns. In point of fact, the muntjac of southeast Asia experiences a postpartum estrus, thus becoming pregnant again as soon as she has delivered her last fawn. Soper (1969) documents the case of a female that produced 4 offspring in the space of 21 months. The white-tailed deer and brockets of South America also reproduce *seriatim* (Brokx, 1972; Gardner, 1971).

In the case of male tropical deer it is a different story. Although they are sexually fertile throughout the year, as is the case with females, they have the problem of how to replace their antlers in the absence of an annual day length cycle (Goss, 1963). By control mechanisms that are not yet understood, these deer cast and regrow their antlers each year as do their temperate zone relatives, but they do not do so in unison. The aseasonal nature of antler replacement cycles is apparently correlated with the lack of a specific fawning season, the implication being that the time of year when a buck replaces his antlers is determined by when he happened to have been born. Longitudinal observations of captive white-tailed deer in Venezuela (T. Blohm, personal communication) confirm that they are somehow able to replace their antlers at precisely the same time each year. How they are aware that 12 months have elapsed in an aseasonal environment is a mystery. Even when such animals live in temperate zone zoos, they still replace their antlers and reproduce at all times of year despite the obvious disadvantages of giving birth sometimes in the middle of winter. Tropical deer have apparently lost the ability, or need, to monitor and react to seasonal changes in day lengths. Whatever may be the mechanism by which these animals measure the passage of each 12-month period, it is apparently independent of latitude and photoperiod.

Still another enigma is the fact that, unlike temperate zone species, tropical males remain fertile all year irrespective of the condition of their antlers. At higher latitudes, males are sterile when they are growing new antlers because of the atrophy of their testes and the cessation of spermatogenesis. However, tropical species continue to produce sperm even while their antlers are in velvet (Goss, 1963). Because the antler replacement cycle in temperate zone species is so dependent on the seasonal fluctuation in testosterone, which is in turn a function of the reproductive condition of the animal, one wonders how the tropical antler cycle may relate to levels of testosterone in animals that are perennially fertile. Unfortunately, serum levels of male sex hormones have not been investigated in such species, nor has the influence of testosterone on the maturation of antlers been tested in tropical species. However, castration prevents the velvet from being shed in the axis deer as it does in nontropical species (Bullier, 1948). It is possible that antler replacement in the tropics might still be correlated with annual changes in testosterone levels even if these fluctuations were not so extreme as those in temperate zone species. This could explain the not uncommon phenomenon whereby tropical species of deer occasionally fail to replace their antlers in certain years, sometimes for as long as 3 consecutive years (Forsyth, 1889; Goss, 1963; Phillips, 1927–1928). However, once they are replaced, casting and regrowth occur on schedule at the same time of year as in previous cycles. Therefore, although the overt indications of the cycle may be lacking, deer nevertheless track the passage of time with characteristic precision.

In view of what is known regarding deer living near the equator where the days and nights are of equal length throughout the year, one might predict that if temperate zone deer were moved to the equator they would give birth at all times of the year and the males would replace their antlers at 12-month intervals. However, for unaccountable reasons, records of temperate zone species living in zoos on or near the equator are conspicuous by their absence; but judging from experimental results, such deer would probably not react as predicted. When sika deer were held for several years under equatorial lighting conditions [i.e., 12 hours of light alternating with 12 hours of darkness (12L/12D)], they did not replace their antlers for indefinite periods (Goss, 1969b). If such deer were introduced to constant and equal photoperiods after the winter solstice, they cast and regrew new antlers the following spring and summer, but not thereafter. Animals started earlier in the autumn did not replace their antlers in the spring, nor for as long as they were held under such constant lighting conditions. One animal failed to grow new antlers for almost 4 consecutive years. Curiously, if weanling fawns are introduced to these simulated equatorial lighting conditions in the early autumn, they nevertheless grow their first sets of antlers the following spring. Even when pregnant does are kept on 12L/12D so that their offspring are exposed from birth to artificial lighting conditions, the males still develop their first antlers on schedule. However, these antlers are not replaced in subsequent years. As pointed out in Chapter 6, a deer's first antlers are programmed independently of the photoperiod.

Unusual light schedules can sometimes cause unexpected results. Some animals, for example, cast their antlers but did not replace them. There remained areas of exposed bone on the pedicles which failed to heal (Fig. 112). Another animal lost one antler and grew a short replacement, but this occurred only on one side, the other antler remaining bony (Fig. 113). Castration of deer held at 12L/12D results in the loss of both antlers several weeks later, followed by the growth of new ones. Apparently, when sika deer are held on constant and equal day lengths for indefinite periods they remain permanently in the fertile phase of their reproductive cycle. The testes are enlarged and spermatogenic. The coat remains in the dark unspotted winter condition. Therefore, one might predict that if such deer were in fact introduced to a zoo on the equator they would reproduce at more frequent intervals throughout the year as do the native tropical species, but the males would keep the same set of antlers throughout life. There would probably be some molting problems, too. The apparent absence of temperate zone deer in tropical zoos may be attributable more to the lack of photoperiodic stimulation than to problems of thermal adaptation.

Circannual Rhythms

The foregoing experiments suggest that deer are entirely dependent on fluctuations in day length as the stimulus for antler replacement. However, the story is more complicated. When the length of day is held constant but is unequal to the length of night, deer are then able to replace their antlers from time to time. In experiments in which they were maintained on 8L/16D, 16L/8D, or 24L/OD, groups of sika deer tended to cast and regrow their antlers at irregular intervals (Goss, 1969b). The animals did not respond in synchrony with each other, but the average duration of their antler replacement cycles was about 10 months. This proved that it is not necessarily the change in day length that is required for antler renewal, but that as long as the days and nights are unequal deer are able to express a rhythm of antler replacement governed by some innate timing mechanism that approximates a year. In the case of daily rhythms, animals are still capable of cycling even when not held under alternating light and dark periods. Such cycles are usually a little shorter or longer than 24 hours, and are therefore referred to as circadian rhythms (circa, about; diem, day). By analogy, various biological cycles that approximate a year under certain artificial lighting conditions are referred to as circannual rhythms (Pengelley, 1974). Why it is that antler rhythms are expressed when the ratio of light to dark does not equal one, but are suppressed when the days and nights are constant and equal, is not understood. Other circannual systems persist on 12L/12D. Nevertheless, one can ask just how unequal the days and nights must be for circannual rhythms to be expressed. Accordingly, groups of sika deer were held under various lighting conditions differing progressively from 12L/12D (Goss, 1977). Antler replacement failed to occur in such deer exposed to 12

¼L/11¾D, 12½L/11½D, and 12¾L/11¼D. However, of 4 deer kept at 13L/11D, 3 replaced their antlers. This means that the difference in length between the light and dark periods must be in the order of 1.5–2 hours for the circannual rhythm of antler replacement to cut in (Fig. 135). If these differences are plotted in terms of day lengths at the solstice as a function of latitude (Fig. 136) one can bracket a zone on either side of the equator approximating the region of transition between seasonally breeding deer in the temperate zone and aseasonal ones in the tropics (Fig. 137). This zone is

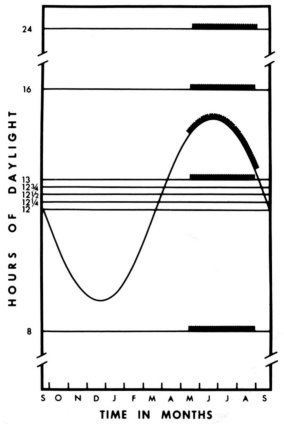

Fig. 135. Summary of experiments on the reactions of antler growth cycles to constant day lengths of different durations. The sine curve represents the natural changes in day lengths throughout the year at about 42° N latitude. Antlers are normally in velvet from early May to early September. Yearling sika deer held under conditions of 8L/16D, 13L/11D, 16L/8D, and 24L/OD since the autumnal equinox all shed and regrew antlers in the spring in synchrony with those outdoors. Animals exposed to light for 12, 12¼, 12½, and 12¾ hours per day, however, failed to replace their antlers the following year. (From Goss et al., 1974.)

probably at about 14–18° latitude (Goss et al., 1974), but it would be useful to have specific confirmation of this. Central and South America are the only places in the world where there is a continuous land bridge across the equator inhabited by a single kind of deer. One can only hope that data on the life cycles of Latin American deer will be forthcoming before they become extinct through destruction of their habitats.

There remains the intriguing problem of why antlers are not replaced when deer are held at 12L/12D, but are when the photoperiod is 8, 13, 16, or 24 hours per day. The failure of the circannual rhythm to express itself on 12L/12D could be attributed either to the parity of the light and dark periods under these conditions, or to the fact that the light (or dark?) period is 12 hours long. To explore these alternatives, sika deer have been kept at 6L/6D, a lighting regime in which the light and dark periods are still equal to each other, but neither is 12 hours long. Under these conditions, deer do replace

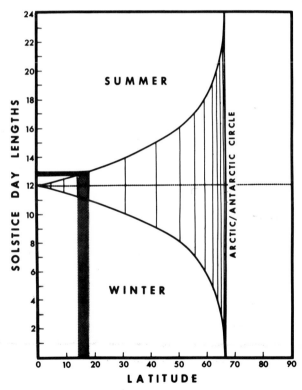

Fig. 136.   Graph of the maximum (summer) and minimum (winter) lengths of natural days as a function of latitude. Shaded areas indicate the zone between 14° and 18° N latitude where the length of day at the summer solstice is between 12¾ and 13 hours. (From Goss et al., 1974.)

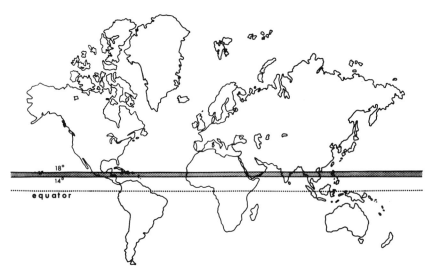

Fig. 137. The relation of 14° and 18° N latitudes to the world's land masses. Only in Central and South America is there a land bridge between the hemispheres inhabited by deer. The shaded zone is believed to approximate the transitional region between the seasonal pattern of reproduction in the temperate zone and aseasonal reproduction in the tropics. (From Goss *et al.*, 1974.)

their antlers in the spring (Goss, 1977). However, the problem with this experiment is that the total light and dark periods still total 12 hours apiece per day. To avoid this problem, other groups of sika deer have been held on days totaling 10 hours (5L/5D), 16 hours (8L/8D), and 42 hours (21L/21D). In these cases, the light and dark periods remain equal, but neither one totals 12 hours in any 24-hour period. Introduced to these lighting conditions at the autumnal equinox in September, the deer tended to cast their antlers in winter or spring, somewhat ahead of schedule. They replaced them again the next winter in accordance with the expression of a circannual rhythm. As in the case of other animals (Gwinner, 1981), deer held on light regimes they cannot follow do not measure the passage of a year simply by counting the days. Therefore, the circannual cycle is not the sum of 365 circadian ones, but must be explained in terms of physiological mechanisms not necessarily related directly to the photoperiod.

References

Brokx, P. A. (1972). Ovarian composition and aspects of the reproductive physiology of Venezuelan white-tailed deer (*Odocoileus virginianus gymnotis*). *J. Mammal.* **53,** 760–773.

Bullier, P. (1948). Curieuse anomalie des bois chez un cerf castré (*Axis axis*). *Mammalia* **22,** 271–274.

Donne, T. E. (1924). "The Game Animals of New Zealand." Murray, London.

Fletcher, T. J. (1974). The timing of reproduction in red deer (*Cervus elaphus*) in relation to latitude. *J. Zool.* **127,** 363–367.

Forsyth, J. (1889). "The Highlands of Central India." Chapman & Hall, London.

Gardner, A. L. (1971). Postpartum estrus in a red brocket deer, *Mazama americana,* from Peru. *J. Mammal.* **52,** 623–624.

Goss, R. J. (1963). The deciduous nature of deer antlers. *In* "Mechanisms of Hard Tissue Destruction" (R. F. Sognnaes, ed.), Publ. No. 75, pp. 339–369. Am. Assoc. Adv. Sci., Washington, D.C.

Goss, R. J. (1969a). Photoperiodic control of antler cycles in deer. I. Phase shift and frequency changes. *J. Exp. Zool.* **170,** 311–324.

Goss, R. J. (1969b). Photoperiodic control of antler cycles in deer. II. Alterations in amplitude. *J. Exp. Zool.* **171,** 223–234.

Goss, R. J. (1976). Photoperiodic control of antler cycles in deer. III. Decreasing versus increasing day lengths. *J. Exp. Zool.* **197,** 307–312.

Goss, R. J. (1977). Photoperiodic control of antler cycles in deer. IV. Effects of constant light:dark ratios on circannual rhythms. *J. Exp. Zool.* **201,** 379–382.

Goss, R. J. (1980). Photoperiodic control of antler cycles in deer. V. Reversed seasons. *J. Exp. Zool.* **211,** 101–105.

Goss, R. J. (1983). Control of deer antler cycles by the photoperiod. *In* "Antler Development in Cervidae" (R. D. Brown, ed.). Caesar Kleberg Wildl. Res. Inst., Kingsville, Texas (in press).

Goss, R. J., Dinsmore, C. E., Grimes, L. N., and Rosen, J. K. (1974). Expression and suppression of the circannual antler growth cycle in deer. *In* "Circannual Clocks, Annual Biological Rhythms" (E. T. Pengelley, ed.), pp. 393–422. Academic Press, New York.

Gwinner, E. (1981). Circannual rhythms: Their dependence on the circadian system. *In* "Biological Clocks in Seasonal Reproductive Cycles" (B. K. Follett and D. E. Follett, eds.), pp. 153–169. Scientechnica, Bristol.

Handley, C. O., Jr. (1950). Game mammals of Guatemala. *U. S., Dep. Inter., Fish Wildl. Serv., Spec. Sci. Rep.* **5,** 141–162.

Jaczewski, Z. (1954). The effect of changes in length of daylight on the growth of antlers in the deer (*Cervus elaphus* L.). *Folia Biol. (Krakow)* **2,** 133–143.

Leader-Williams, N. (1978). The history of the introduced reindeer of South Georgia. *Deer* **4,** 256–261.

Leopold, A. S. (1959). "Wildlife of Mexico. The Game Birds and Mammals." Univ. of California Press, Berkeley.

Mearns, E. A. (1907). Mammals of the Mexican boundary of the United States. *Bull.—U. S. Natl. Mus.* **56.**

Pengelley, E. T., ed. (1974). "Circannual Clocks, Annual Biological Rhythms." Academic Press, New York.

Phillips, W. W. A. (1927–1928). Guide to the mammals of Ceylon. Part VI. Ungulata. *Ceylon J. Sci., Sect. B* **14,** 1–50.

Ransom, A. B. (1966). Breeding seasons of white-tailed deer in Manitoba. *Can. J. Zool.* **44,** 59–62.

Rowan, W. (1925). Relation of light to bird migration and developmental changes. *Nature (London)* **115,** 494–495.

Soper, E. (1969). "Muntjac." Longmans, Green, London.

Internal Influences

Aside from the environmental factors to which deer are adapted and by which their annual cycles are controlled, antlers are subject to a variety of endogenous influences that affect not only their development and seasonality, but also the uses to which they are put. Inheritance has been shown to play a major role in shaping antler morphology, as the series of almost identical antlers grown during the lifetime of an individual stag so convincingly testifies (Fig. 138). Quality and quantitiy of diet are important modulating influences too, but the distinction between nurture and nature is difficult to define. However, the most proximate control mechanisms are the hormones that affect antler growth, of which there are quite a few. It is through endocrine mechanisms that the profound influences of the light cycle are mediated.

Antlers are secondary sex characters. As such, they are primarily affected by the annual fluctuations in the secretion of sex hormones (Short, 1968; Tachezy, 1956). Other hormones may be involved too, but the gonads play a pivotal role in the production of antlers in the first place, and their regeneration each successive year. In order to appreciate the endocrine basis of antler development, it is first necessary to understand the reproductive cycles of deer, cycles with which the recurrent growth of antlers is so intimately coordinated.

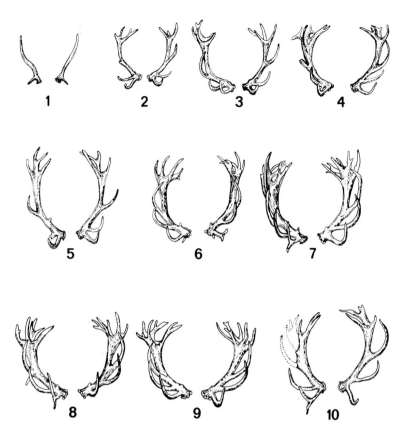

Fig. 138. Antlers produced by a red deer stag in 10 successive years, starting with his first ones grown as a yearling. Similarities from year to year are evidence of genetic factors. Variations presumably indicate the effects of injuries or of systemic influences. (After Millais, 1897.)

Testosterone

If puberty is defined as the onset of spermatogenesis, then most male deer normally reach this stage of maturity in their second year, that is, in the autumn after their first birthday. In a few species, including the muntjac, as well as those deer that sometimes grow their first short sets of antlers as fawns instead of as yearlings (e.g., roe deer, moose, *Odocoileus*, reindeer, and caribou), precocious individuals may undergo limited spermatogenic activity during their first autumn. This adolescent reaction is not to be equated with the degree of sexual maturation (and social status) required for successful mating. Nevertheless, the production of a deer's first antlers cannot occur in the absence of the male sex hormone, the secretion of which depends on sufficient differentiation of secretory cells (Leydig cells) in the testes. As a fawn matures, his testes keep pace with the enlargement of the rest of the body (Chaplin and White, 1972). However, at the onset of puberty the testes enlarge much more rapidly than other organs as their seminiferous tubules begin to produce sperm and their Leydig cells secrete testosterone. Rising levels of this hormone stimulate development of the seminal vesicles as well as the genitalia. However, even before this occurs the earliest signs of antler production become evident (Chapter 6).

Before the antlers themselves can be produced, the pedicles from which they develop must form. The prospective sites of pedicle development are recognizable even before birth as cowlicks overlying the frontal bones. A slight protuberance in the bone may be palpable beneath these cowlicks. When the periosteum in this location is stimulated by sufficient levels of testosterone, it begins to lay down increasing amounts of bone to give rise to an exostosis that pushes upward beneath the overlying scalp. In the absence of testosterone, or in the presence of the female sex hormone, estrogen, this reaction fails to occur and antlers do not develop.

Following this first round of testicular growth, a process that appears to be preprogrammed because it takes place independent of seasonal changes in the environment, the male gonads are thereafter subject to annual cycles of activity which reflect correspondingly dramatic changes in their histology. Like other animals native to the temperate zone, deer undergo annual alternations between fertility and sterility. With gestation periods spanning two to three seasons, the fertile phase usually peaks in autumn, while the infertile one dominates spring and summer. These sexual alternations are accompanied by wide swings in testicular activity (Wislocki, 1949). In the spring the testes are atrophic (Fig. 139). Their narrow seminiferous tubules are lined with quiescent cells no longer engaged in spermatogenesis. In between these tubules are the Leydig cells, the small sizes of which reflect their secretory inactivity (Aughey, 1969). With the arrival of spring, there may be a transient activation of spermatogenesis at about the time when the antlers begin to grow (Markwald et al., 1971; Robinson et al., 1965; West and Nordan, 1976a). However, it is not until early summer that unmistakable signs of fertility first appear (Lambiase et al., 1972; Lincoln, 1971). The testes enlarge and spermatogonia lining the seminiferous tubules embark on the long process of spermatozoan differentiation. The Leydig cells increase in size as they start to manufacture testosterone (Frankenberger, 1954). Even while the antlers are still in velvet the testes prepare for the forthcoming mating season. Indeed, it is the rising levels of testosterone that add the finishing touches to the developing antlers. Elongation slows down, ossifica-

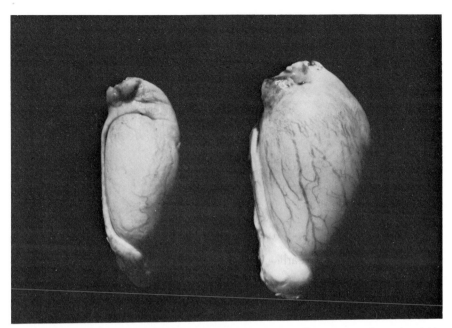

Fig. 139. Testes from sika deer of comparable body sizes illustrating an almost threefold difference in gonadal weight between the spring (left) and fall (right).

tion strengthens the inner tissues by the deposition of mineral salts, and in due course the velvet is shed. Therefore, by early fall deer are equipped fore and aft with the wherewithal for the coming rutting season. Their testes (Fig. 139) have tripled or quadrupled in mass (Chapman and Chapman, 1970; Hochereau-de Reviers and Lincoln, 1978), the seminiferous tubules have doubled their diameters (Lincoln, 1971), and the rising production of testosterone floods the bloodstream with concentrations manyfold greater than earlier in the year (West and Nordan, 1976a; Whitehead and McEwan, 1973; Whitehead and West, 1977). This endocrine crescendo reaches a peak in October or November for most species of deer (Lincoln, 1971; Lincoln and Kay, 1979; McMillan et al., 1974), a time of year when feverish reproductive activity dominates the lives of the stags. Does are not left in peace until they have been bred.

After the rutting season there is a decrease in the production of testosterone (Fig. 140). It is this decline that is responsible for the casting of the antlers which may occur in early winter in *Odocoileus,* moose, reindeer and caribou, but not until spring in most other species (e.g., wapiti, red deer, sika, fallow). But there are exceptions. In the case of the roe deer which

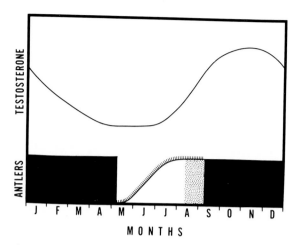

Fig. 140.   Relative levels of serum testosterone compared with the antler cycle at different times of year. Vertical cross-hatching indicates time in velvet. Stippling is period of mineralization. Black is phase of bony antlers.

breeds in the summer, the decline in testicular activity, including testosterone output, takes place as early as August, reaching minimal levels in the autumn (Sempéré and Boissin, 1981, 1983). Ordinarily one would expect roe antlers to be lost earlier in the fall than is normally the case (i.e., November or December). Possibly the secondary rut that has been reported sometimes to occur in the autumn (Stieve, 1950) may be associated with the prolonged retention of antlers by this species. Indeed, there are rare cases in which roe bucks have been reported to grow two sets of antlers a year (Prior, 1968; Ullrich, 1961), which might be predicted if testosterone levels dropped precipitously enough following the summer rut to trigger casting and regrowth of antlers in the late summer or early fall. Although testicular parameters (Short and Mann, 1966) and testosterone levels (Sempéré and Boissin, 1981, 1983; Sempéré et al., 1980) do in fact decline sharply after the summer rut, roe bucks normally do not exhibit secondary mating periods or antler cycles. Such anomalies probably may be explained in other ways.

   In still other species, notably those native to tropical climates (Chapter 11), spermatogenesis may go on all year around, but antlers are cast and regrown annually. One can only presume that this paradox may be explained by yearly fluctuations in testosterone production great enough to effect the loss and replacement of antlers but not sufficient to interfere with fertility. Indeed, the occasional retention of the same set of antlers for 2 years or more by such animals may reflect the fact that hormone levels may not always dip low enough each year to initiate a new round of antler production. Data on testosterone levels in tropical deer, assayed at different times of year in relation to the antler cycle, are greatly desired.

Testosterone does not affect fawns and adults alike. In the former, it is necessary for the production of pedicles from the frontal bones. Because testosterone is antagonistic to antler development in adults, one wonders how the first antlers are produced by a fawn or yearling if testosterone is required for their initiation. Sempéré and Boissin (1983) have documented the rise in plasma testosterone of male roe fawns during their first summer and fall when the initial "button" antlers grow, not declining until the velvet is shed in January. This aspect of the problem is in need of clarification.

In older deer, which replace their antlers every year, testosterone is not necessary for the onset of antler growth, but is required for their maturation (Fig. 140). Although initiation of antler bud formation often coincides with, or even anticipates, the casting of the old antlers, these two events may be separated by up to several months in those species that lose their old antlers precociously in the early winter. In either case, it would appear to be the low level of testosterone in the late winter and early spring that is somehow responsible for promoting antler growth. In fact, it has been shown that if testosterone is administered to deer before their old antlers have dropped off, casting may be postponed for as long as hormone injections continue (Goss, 1968; Jaczewski and Galka, 1967). However, if a deer is treated with testosterone while it is in velvet, his antlers soon stop growing, internal ossification occurs, and the velvet is shed prematurely (Fig. 141). Jaczewski and Michalakowa (1974) showed that human chorionic gonadotrophin interferes with the casting and regrowth of red deer antlers, probably by stimulating testosterone secretion.

It is perhaps logical to conclude that antlers are one of the target organs of

Fig. 141. Injections of testosterone while sika antlers are in velvet (a) causes cessation of growth and shedding of velvet, resulting in the precocious maturation of the antlers (b). (From Goss, 1968. Reprinted by permission from *Nature*, **220**, No. 5162, pp. 83–85. Copyright © 1968 Macmillan Journals Limited.)

testosterone. Bubenik *et al.* (1974a) have shown by immunofluorescent techniques that it occurs in the differentiating cartilage cells near the growing tip of an antler as well as in association with hair follicles in the velvet. Implantation of testosterone pellets beneath the skin of growing antlers, however, does not bring about a localized maturation as might have been expected if the tissues of the antler were directly susceptible to its effects. Further, it remains to be explained why testosterone should be localized in the velvet antler, the growth of which is actually promoted by the lack of this hormone. Therefore, the possibility remains that testosterone may act indirectly on antlers via some other agent as yet to be identified.

Estrogen

Females do not grow antlers, but the distaff side of reproduction is not unrelated to the production and replacement of antlers. Like the males, does alternate annually between fertility and infertility. During the mating season the females come into heat every 3–4 weeks in *Odocoileus* (Thomas and Cowan, 1975) and the moose (Peterson, 1974), 15–20 days in red deer (Guinness *et al.,* 1971), and 10–12 days in reindeer and caribou (McEwan and Whitehead, 1972). Ovulation occurs at estrus, which may last about 24 hours and is the only time when the female is receptive to the male. If a doe does not become pregnant the first time around, she usually will the next. In most species, a doe will continue her estrous cycles throughout the fall and into the winter if she is not bred.

Assuming conception occurs, the fertilized egg undergoes early embryonic development within the fallopian tubes and uterine cavity. In the meantime, the Graafian follicle from which the egg was produced transforms into a corpus luteum, a transient endocrine organ which develops from former follicle cells now capable of secreting progesterone. This hormone is necessary to prepare the uterine wall for implantation by the embryo, an event normally occurring about 1 week after fertilization (except in the case of delayed implantation in the roe deer). Accessory corpora lutea have been identified in the ovaries, sometimes in the same one from which the egg was produced, sometimes in the opposite ovary (Douglas, 1966; Harder and Moorehead, 1980). These accessory corpora lutea are believed to supplement progesterone secretion by the primary one to help maintain pregnancy throughout the long period of gestation. In certain tropical species of deer, such as the muntjac of Asia as well as the brockets and white-tailed deer of South America, the females experience a postpartum estrus as a result of which they may become pregnant again soon after giving birth (Chapter 2). In such instances, the reproductive rate is accelerated because a deer need not wait until she has weaned one fawn before conceiving another. However, in nontropical species, where the reproductive cycle is closely coordinated with the seasons, lactation and pregnancy do not overlap.

In nonpregnant white-tailed does the levels of estrogen secreted by the ovaries do not vary significantly throughout the year (Plotka et al., 1977a). However, in pregnant females estrogen levels in the serum begin to rise several months before birth, climbing steeply in the last 6 weeks and dropping rapidly after delivery (Plotka et al., 1977b). It is believed that this abrupt decline in estrogen levels may be responsible for the casting of antlers by female reindeer and caribou at about the time of parturition (Lent, 1965). Presumably the elevated estrogen in the pregnant female causes her antlers to be retained, whereas those of barren reindeer and caribou are cast earlier in the season when their estrous cycles cease and estrogen levels drop.

This interpretation is consistent with the experimental evidence of the effects of estrogen injections on deer. When administered to castrated male red deer in January, a time of year when they do not normally exhibit rutting behavior, they begin to roar and fail to cast their antlers as long as the hormone is administered (Fletcher and Short, 1974). When given in August, the antlers are induced to shed their velvet prematurely. Comparable experiments on the sika deer have confirmed that estrogen prevents or postpones loss of the old antlers, and promotes shedding of the velvet if given to animals with growing antlers (Goss, 1968). Therefore, estrogen exerts effects very similar to those of testosterone on adult males, despite the fact that these two hormones have opposite effects on fawns.

One reason females do not ordinarily grow antlers is that as fawns they produce no pedicles. Removal of the ovaries alone is not sufficient to induce pedicle development in prepubertal females, but if such animals are also treated with testosterone, pedicles will develop. The production of antlers from these pedicles depends on the trauma of amputating their ends (Chapter 14). In the case of males, although castration as fawns prevents pedicle and antler development, the administration of estrogen to otherwise intact male fawns will likewise interfere with the production of pedicles. Therefore, pedicle development is promoted by testosterone and inhibited by estrogen. Antler growth is inhibited by both hormones.

Progesterone

As indicated earlier, the ovary secretes progesterone as well as estrogen. Not surprisingly, this hormone is not found in the blood of males, nor of female fawns (Sempéré, 1977). In nonpregnant white-tailed does, its lowest levels are found in summer and fall, and while maximal concentrations occur in winter (Plotka et al., 1977a). In pregnant females, progesterone levels remain elevated throughout gestation, when they are five or more times higher than in nonpregnant animals (Plotka et al., 1977b). At birth, the levels of progesterone in the mother are abruptly decreased owing to the involution of her corpora lutea as well as the loss of the placenta, which is itself a source of progesterone.

The role of progesterone in delayed implantation in roe deer is of interest. It has been confirmed (Sempéré, 1977; Short and Hay, 1966) that the corpus luteum is active during the period of delayed implantation. However, once embryonic development resumes in early winter, the levels of progesterone in the blood in roe deer rise. This would appear to be a consequence rather than a cause of implantation.

Pituitary Hormones

The gonads are under the control of the pituitary. Weights of this gland have been found to fluctuate between a minimum in the winter and a maximum in late spring (Hoffman and Robinson, 1966; Lincoln, 1971). Acidophilic cells, which secrete growth hormone, become enlarged with increased secretory activity in the spring, but decrease as the summer progresses to minimal dimensions in the winter (Nicolls, 1971; Stošić and Pantić, 1966; West and Nordan, 1976b). Their numbers are also depleted in the fall and winter. Cell types responsible for the secretion of other hormones by the pituitary have also been found to undergo seasonal variations (Freund, 1955). It is conceivable that the pituitary might secrete a hormone directly responsible for initiating antler growth, an event otherwise attributed only to the absence of testosterone. The identification of such a hypothetical hormone has for many years been one of the goals of zoologists interested in the endocrinology of antler growth. They are still searching.

Some years ago researchers at the Massachusetts General Hospital in Boston attempted to hypophysectomize some white-tailed deer (Hall et al., 1966). Accordingly, 6-month-old males, properly sedated, were brought to the hospital for the operation. Although not all such operations were successful, one animal survived this radical surgery for up to 13 months, with the aid of ACTH and cortisone replacement therapy. This deer underwent little or no growth subsequent to hypophysectomy, nor did he molt regularly. As expected, he failed to grow antlers, except for the very slight production of pedicles. Administration of large doses of bovine growth hormone had no apparent effect on antler production. Therefore, it would seem that the pituitary is indispensable for the development of pedicles and antlers, although the possibility remains that this could be attributed to the lack of gonadotrophic hormones and testosterone, upon which pedicle development is known to depend.

It would seem logical that growth hormone might be a likely candidate for the hypothetical "antlerotrophic" hormone. The pituitary cells responsible for its secretion are larger and contain more granules in the spring when antler growth begins in the red deer (Stošić and Pantić, 1966). The levels of growth hormone in the plasma of white-tailed deer are high in the spring, and decline gradually throughout the period of antler growth and rut (Bahnak et al., 1981; Bubenik et al., 1974b, 1975), but Schulte et al. (1980b, 1981b) found no seasonal changes in the pituitary cells that secrete growth hormone. Conflicting data on growth hormone levels in the blood of reindeer in Norway and Svalbard have also been reported by Ringberg (1979). Thus, the role of growth hormone in stimulating antler development must remain little more than an interesting possibility pending the availability of more convincing evidence.

Another hormone that could be involved is prolactin. The levels of this hormone in the blood of white-tailed deer of both sexes reach a peak in the spring or summer, and drop to minimal levels in winter (Mirarchi et al., 1978; Schulte et al., 1980a,b), a seasonal variation consistent with the cycle of antler production as well as the size of prolactin-secreting cells in the pituitary (Schulte et al., 1981b). However, implants of this hormone into castrated red deer stags failed to promote antler growth (Jaczewski and Krzywinska, 1974). Again, it would seem prudent to withhold judgment on the possible role of prolactin in antler growth until more data are forthcoming.

Luteinizing hormone (LH) is a gonadotrophin secreted by pituitary cells which increase in number toward the end of the summer (Stošić and Pantić, 1966). When the mating season approaches, levels of this hormone in the blood rise in connection with the role of the photoperiod in promoting ovulation in the female and testosterone secretion in the male (Bubenik et al., 1982; Lincoln and Kay, 1979; Mirarchi et al., 1978). This hormone is not known to have a direct effect on the antlers.

Follicle stimulating hormone (FSH) is another gonadotrophic hormone which is secreted maximally during the rutting period. Its production by the pituitary falls to minimal levels in the spring (Bubenik et al., 1982; Mirarchi et al., 1978). Again, its target organ is the gonads, not the antlers themselves.

Adrenal Steroids

The adrenal glands secrete two classes of hormones, one from the outermost zone (zona glomerulosa) to maintain the balance of electrolytes in the body fluids, the other produced by the inner zone of the adrenal cortex (zona fasciculata) responsible for regulating glucose levels in the blood. The adrenal cortex has been reported to be smaller in the winter than in the fall or spring (Hoffman and Robinson, 1966; Lincoln, 1971). Cortisol levels have been recorded as unchanged year round (Bubenik et al., 1974b, 1975a; Ringberg, 1979), or to be maximal in spring and minimal in fall (Bubenik et al., 1977). Adrenal cortical hypertrophy may occur in deer under conditions of stress (Hughes and Mall, 1958), a reaction which may delay antler casting (Topinski, 1975). The relation between the adrenal cortex and antler growth clearly deserves further attention.

Thyroid Hormone

The thyroid gland secretes thyroxin, a hormone with basic effects on metabolism and therefore physiologically related to numerous functions of the body. The activity of the thyroid has been investigated in a number of different types of deer to determine how it varies from one season to another. Not all such investigations have shown seasonal changes (Grafflin, 1942; Lincoln, 1971), but those which have report increased secretory activity in the spring or summer compared with the winter (Bahnak et al., 1981; Bubenik and Bubenik, 1978; Hoffman and Robinson, 1966; Ringberg, 1979; Ringberg et al., 1978). Although this could be taken to indicate a significant effect on antler growth, it might also be nothing more than a response to seasonal conditions unrelated to antler growth and reproductive cycles. In fact, in the roe deer the height of the thyroid epithelium (one indicator of thyroid activity) has been shown to increase in the spring and summer and decrease thereafter (Pantić and Stošić, 1966), despite the fact that this animal grows his antlers in the winter and has a rutting season in the summer. Therefore, it may be premature to jump to the conclusion that the ups and downs of thyroxine secretion are causally related to antler growth. Nevertheless, it has been claimed that injections of thyroxine promote elongation of antlers in roe deer (Lebedinsky, 1939), reindeer, and fallow deer (Bruhin, 1953).

The Effects of Pinealectomy

Experiments on laboratory rodents have confirmed that the pineal gland plays a major role in coordinating cycles of reproductive activity with photoperiodic rhythms. The pineal secretes melatonin in the dark, but not in the light. Melatonin prevents the pituitary from secreting gonadotrophins, without which the gonads cannot grow and function. Pinealectomy therefore causes gonadal hypertrophy.

This scenario makes sense for small mammals and birds because their short periods of gestation or incubation permit breeding to occur in the same season as birth or hatching. Such animals become reproductively active as the days lengthen in the spring.

Larger mammals have longer gestations and must therefore breed while the days grow shorter if births are to occur in the spring as they must. If the pineal regulates such cycles, it cannot do so in the same way as in rodents. Some step in the chain of events must be reversed. Melatonin could be secreted in the day instead of the night; or it could stimulate rather than inhibit secretion of gonadotrophins from the pituitary. It may not even be involved at all with photoperiodically governed cycles. Deer are excellent animals in which to explore the possible role of the pineal gland in controlling reproductive cycles in short-day breeders because their antlers are such accurate indicators of their reproductive condition.

Once surgical techniques were developed for pinealectomy in deer (Brown et al., 1978; Letellier et al., 1978), it was possible to discover once and for all if the pineal gland was as important for deer as rodents. The results have not been clear-cut.

Without the pineal, white-tailed deer still express annual rhythms, contrary to what might have been expected by extrapolation from smaller mammals. However, such rhythms tend to be delayed, as the investigations of Plotka et al. (1979) have shown. Instead of casting their antlers in winter, pinealectomized white-tails lose them in the spring. The period of antler growth, normally from May to September, is postponed until about June through late fall when the velvet is finally shed. The antlers that are produced are abnormally small and incompletely formed.

The pelage also shows the effects of pinealectomy. The winter coat, ordinarily molted in spring, may be retained through July. The fall molt is correspondingly delayed, and sometimes the dark winter and light summer coats may be mixed in the same individual.

The reproductive cycle itself, like those of antler and hair replacement, tends to be delayed. Testicular development occurs later than normal, but the testes become enlarged as they do in pinealectomized rodents. The delayed rise in testosterone levels is matched by the buck's belated expression of aggressive behavior. Prolactin levels are lowered (Schulte et al., 1981a).

Inasmuch as the pineal is responsive to day lengths, blindness should be tantamount to pinealectomy. In the 1960s, a blind wapiti was held under observation at Colorado State University for a number of years (R. W. Davis, personal communication). This animal had been parasitized by *Elaeophora schneideri*, a nematode worm that infects the retina and optic nerve. Although one cannot be certain that he was totally blind, his behavior strongly suggested it. Nevertheless, this wapiti replaced his antlers on schedule each year, and sired a number of offspring. It is possible that his pineal gland could have reacted directly to incident light penetrating the skull in weak but detectable intensities. Indeed, the fact that this animal's annual cycles were not disrupted by the lack of sight indicates that blindness is not entirely equivalent to pinealectomy in this species.

These findings suggest that the control of biological rhythms may be more complex than heretofore believed. The pineal gland clearly plays an important role in translating seasonal changes in the photoperiod into developmental and physiological responses. However, the persistence of these cycles, albeit delayed, in the absence of the pineal gland indicates the operation of other factors in regulating seasonal rhythms. Circannual rhythms (Chapter 11) suggest themselves as possible explanations of the continued but irregular cycles of molting, antler replacement and reproductive activity expressed by pinealectomized deer. However, if circannual cycles are not of pineal origin, it is not clear where they are generated.

Nutrition versus Genetics

As seasonal breeders, deer in temperate zones exhibit significant fluctuations in how much they eat at different times of year. Whether grazers or browsers, males consume great quantities during the summer months, providing minerals for antler growth and allowing storage of fat reserves for the coming fall and winter. In the rutting season, they are far too preoccupied with their sexual obligations to take much time to eat. After rut, males are often emaciated and exhausted at a time of year when environmental conditions worsen. A harsh winter takes its toll of vulnerable deer.

Interest in antlers as hunting trophies has prompted a number of investigations of how antler size might be affected by diet and inheritance. Observations of wild mule deer referred to by Anderson (1981) have shown that the percentage of spike bucks in a population (indicating poor antler growth), as opposed to those with branched antlers, increases in years with lowered rainfall and decreased availability of food. Such deer not only have smaller antlers, but are generally in poorer condition. In wet years there are fewer spike bucks and larger numbers of yearlings with forked antlers. Malnutrition during the first year of life not only delays puberty (Pimlott, 1959), but also defers the onset of pedicle and antler development.

Controlled studies by Suttie (1980) on captive animals have supported the role of nutrition in governing the growth of deer and their antlers. For example, red deer in Scotland have been known for many years to be considerably smaller than their continental cousins. To determine if the cause of this difference were hereditary or nutritional, Scottish deer were fed *ad libitum* and found to grow as large as those on the continent under these optimal conditions. Thus, the relatively poor nutrition available in the Scottish highlands is probably a major factor in explaining the smaller dimensions of these deer. Lincoln *et al.* (1971) and Suttie (1983) noted that malnourished red deer carried poorer antlers. The absence of antlers altogether in hummels may be similarly explained (Lincoln *et al.*, 1976).

Most investigations on captive animals have resorted to inadequate diets to determine what effect this might have on growth and development. In white-tailed females, low nutrition decreases ovulation, resulting in lower birth rates, later birth dates, and fewer fawns per doe (Verme, 1965). However, overeating has been shown to repress breeding to the extent that deer made to become obese tend not to produce fawns. Bahnak *et al.* (1981) have noted that when white-tailed does are fed semistarvation diets their thyroxine levels decline in winter, suggesting that this may be a hormonal adaptation to diminished metabolic needs when energy resources are not adequate.

In males, dietary factors are sometimes suspected to cause testicular atrophy. A population of white-tailed deer in Texas studied by Taylor *et al.* (1964) included an unusually high percentage of bucks with atrophic testes and autumn antlers in velvet. Whether this is explained by a local deficiency of an essential mineral or the toxic effects of something the deer may have consumed is not clear. In rats, Johnson *et al.* (1970) have shown that testicular degeneration can be caused by excess cadmium, an effect prevented by selenium or zinc. In such cases, the typical castration syndrome ensues (Chapter 13).

Lathyrism is a condition caused by the consumption of too many sweet peas (*Lathyrus odoratus*). The active ingredient, β-aminopropionitrile (BAPN), interferes with cross-linking between collagen filaments synthesized in its presence. Lathyrism in man is characterized by weakened arterial walls, leading to aortic aneurysms. In developing animals, limb malformations may occur. In view of the high collagen content of growing antlers, it might be expected that BAPN would be responsible for morphogenetic abnormalities in antlers growing under its influence. When sika deer were fed a diet of increasing concentrations of sweet pea seeds (up to 50%) while growing antlers, they exhibited typically loose joints and eventually had difficulty in standing. Nevertheless, their growing antlers continued to develop almost normally. Evidently the role of collagen in shaping an antler must be compensated by the deposition of cartilage and bone that lend sufficient strength to the antler to sustain its normal morphology.

Studies of white-tailed deer in Pennsylvania have shown that undernutrition (i.e., low calcium, phosphorus, or protein) not only results in decreased body weights, but causes such deer to grow abnormally short antlers with fewer points than in animals fed adequate diets (French et al., 1956; Magruder et al., 1957). When calcium and/or phosphorus supplements are added to the diet, antler development is improved (French et al., 1955). The Duke of Bedford (1951) noted that the unusually severe winter of 1946–1947 in England caused a decrease in the calcium content of the grass which his herd of Père David's deer consumed at Woburn Abbey. This calcium-deficient diet was presumably responsible for the production of the "rotten antlers and queer freaks and abnormalities," as he described them. More recently, Wika (1980) reported that when reindeer were fed a poor diet, their body weights were a little more than one-half normal, the coats were in poor condition, and the antlers grew to only one-quarter to one-third the length of control antlers.

There is ample evidence to confirm that antler growth is profoundly influenced by nutritional factors. Their lengths and mass, and therefore the number of points, are reduced in poorly nourished deer. However, the bone density of such stunted antlers has been shown not to be affected by diet (French et al., 1955). In white-tailed deer, the initiation of antler growth in the spring is delayed under conditions of inadequate nutrition but their antlers tend to be cast earlier (French et al., 1955; Hawkins et al., 1968; Plotka et al., 1979). Malnourished red deer may cast their antlers earlier (MacNally, 1976) or later than normal (Suttie, 1983). However, reindeer have been shown to drop their antlers at later dates under conditions of malnutrition, a finding that has prompted the suggestion that the delayed loss of antlers by pregnant reindeer may be attributable not so much to altered hormonal conditions as to the malnutrition of the mothers in competition with their fetuses for sparsely available sustenance (Wika, 1980).

The basic question of whether antler size is determined by diet or heredity has been put to experimental test by Harmel (1982). In a long-range series of controlled studies on Texas white-tailed deer, yearlings with relatively large branched antlers were compared with those carrying only spikes. The inbred progeny of these two lines were fed identical optimal diets ad libitum to learn if the original discrepancies in parental antlers would disappear in the absence of nutritional differences. They did not. The superior buck sired only 5% spike-antlered yearlings in the F_1 generation, and none in backcrosses. The spike bucks produced 44% spike-antlered yearlings in the F_1, and 59% when these were backcrossed. Other antler parameters, as well as body size, were similarly affected. These results underscore the overriding importance of genetics in determining antler quality, and argue in favor of legislation permitting hunters to shoot spike bucks if the deer herd is to be improved.

This is not to say that nutritional factors do not also affect antler size. Ullrey (1983) has shown that underfed white-tailed deer grew smaller and narrower antlers than controls, and that low protein consumption results in reduced numbers of points. Therefore, it is clear that antlers are responsive to dietary as well as genetic influences, but how these two interact is in need of further investigation.

As important as nutrition may be for the full expression of an antler's developmental potential, this is only a modulating influence that determines not whether antlers shall grow or not, but the sizes they are to attain. It has been claimed that genetics determines the form of antlers while nutrition dictates their size. This may be true, except that when the mass is diminished the number of points is likewise decreased. No matter how inadequate the diet may be, it is remarkable that antler growth seems never to be prevented altogether. With the possible exception of hummels, deer in the poorest of physical condition, whether caused by disease or malnutrition, will nevertheless sprout new antlers, albeit diminutive ones, when the time comes. Lower vertebrates exhibit a similar reaction, insisting on replacing lost parts even when dying of starvation. The urge to regenerate must be very strong indeed to be expressed when animals can least afford it.

References

Anderson, A. E. (1981). Morphological and physiological characteristics. In "Mule and Black-tailed Deer of North America" (O. C. Wallmo, ed.), pp. 27–97. Univ. of Nebraska Press, Lincoln.

Aughey, E. (1969). Histology and histochemistry of the male accessory glands of the red deer, (*Cervus elaphus* L.). *J. Reprod. Fertil.* **18**, 399–407.

Bahnak, B. R., Holland, J. C., Verme, L. J., and Ozoga, J. J. (1981). Seasonal and nutritional influences on growth hormone and thyroid activity in white-tailed deer. *J. Wildl. Manage.* **45**, 140–147.

Bedford, Duke of (1951). Père David's deer: The history of the Woburn herd. *Proc. Zool. Soc. London* **121** (Part II), 327–333.

Brown, R. D., Cowan, R. L., and Kavanaugh, J. F. (1978). Effect of pinealectomy on seasonal androgen titers, antler growth and feed intake in white-tailed deer. *J. Anim. Sci.* **47**, 435–440.

Bruhin, H. (1953). Zur Biologie der Stirnaufsätze bei Huftieren. Teil I. *Physiol. Comp. Oecol.* **3**, 63–127.

Bubenik, G. A., and Bubenik, A. B. (1978). Thyroxine levels in male and female white-tailed deer (*Odocoileus virginianus*). *Can. J. Physiol. Pharmacol.* **56**, 945–949.

Bubenik, G. A., Brown, G. M., Bubenik, A. B., and Grota, L. J. (1974a). Immunohistological localization of testosterone in the growing antler of the white-tailed deer (*Odocoileus virginianus*). *Calcif. Tissue Res.* **14**, 121–130.

Bubenik, G. A., Brown, G. M., Wilson, D., Bubenik, A. B., and Trenkle, A. (1974b). GH and cortisol levels in white-tailed deer through antlerogenesis. *Proc. Can. Fed. Biol. Soc.* **17,** 174.

Bubenik, G. A., Bubenik, A. B., Brown, G. M., Trenkle, A., and Wilson, D. A. (1975). Growth hormone and cortisol levels in the annual cycle of white-tailed deer (*Odocoileus virginianus*). *Can. J. Physiol. Pharmacol.* **53,** 787–792.

Bubenik, G. A., Bubenik, A. B., Trenkle, A., Sirek, A., Wilson, D. A., and Brown, G. M. (1977). Short-term changes in plasma concentration of cortisol, growth hormone and insulin during the annual cycle of a male white-tailed deer (*Odocoileus virginianus*). *Comp. Biochem. Physiol.* **58,** 387–391.

Bubenik, G. A., Morris, J. M., Schams, D., and Claus, A. (1982). Photoperiodicity and circannual levels of LH, FSH, and testosterone in normal and castrated male, white-tailed deer. *Can. J. Physiol. Pharmacol.* **60,** 788–793.

Chaplin, R. E., and White, R. W. G. (1972). The influence of age and season on the activity of the testes and epididymides of the fallow deer, *Dama dama. J. Reprod. Fertil.* **30,** 361–369.

Chapman, D. I., and Chapman, N. G. (1970). Preliminary observations on the reproductive cycle of male fallow deer (*Dama dama* L.). *J. Reprod. Fertil.* **21,** 1–8.

Douglas, M. J. W. (1966). Occurrence of accessory corpora lutea in red deer, *Cervus elaphus. J. Mammal.* **47,** 152–153.

Fletcher, T. J., and Short, R. V. (1974). Restoration of libido in castrated red deer stag (*Cervus elaphus*) with oestradiol-17β. *Nature (London)* **248,** 616–618.

Frankenberger, Z. (1954). Interstitiální bunky jelena (*Cervus elaphus* L.). *Cesk. Morfol.* **2,** 36–41.

French, C. E., McEwen, L. C., Magruder, N. D., Ingram, R. H., and Swift, R. W. (1955). Nutritional requirements of white-tailed deer for growth and antler development. *Bull.— Pa., Agric. Exp. Stn.* **600P,** 1–8.

French, C. E., McEwen, L. C., Magruder, N. D., Ingram, R. H., and Swift, R. W. (1956). Nutritional requirements for growth and antler development in the white-tailed deer. *J. Wildl. Manage.* **20,** 221–232.

Freund, J. (1955). The hypophysis of the red deer and its relationship with the reproductive cycle. *Cesk. Morfol.* **3,** 212–221.

Goss, R. J. (1968). Inhibition of growth and shedding of antlers by sex hormones. *Nature (London)* **220,** 83–85.

Grafflin, A. L. (1942). A study of the thyroid gland in specimens of Virginia deer taken at intervals throughout the year. *J. Morphol.* **70,** 21–40.

Guinness, F., Lincoln, G. A., and Short, R. V. (1971). The reproductive cycle of the female red deer, *Cervus elaphus* L. *J. Reprod. Fertil.* **27,** 427–438.

Hall, T. C., Ganong, W. F., and Taft, E. B. (1966). Hypophysectomy in the Virginia deer; technique and physiologic consequences. *Growth* **30,** 383–392.

Harder, J. D., and Moorehead, D. L. (1980). Development of corpora lutea and plasma progesterone levels associated with the onset of the breeding season in white-tailed deer (*Odocoileus virginianus*). *Biol. Reprod.* **22,** 185–191.

Harmel, D. (1983). The effects of genetics on antler quality in white-tailed deer (*Odocoileus virginianus*). *In* ''Antler Development in Cervidae'' (R. D. Brown, ed.). Caesar Kleberg Wildl. Res. Inst., Kingsville, Texas (in press).

Hawkins, R. E., Schwegman, J. E., Autry, D. C., and Klimstra, W. D. (1968). Antler development and loss for southern Illinois white-tailed deer. *J. Mammal.* **49,** 522–523.

Hochereau-de Reviers, M. T., and Lincoln, G. A. (1978). Seasonal variation in the histology of the testis of the red deer, *Cervus elaphus. J. Reprod. Fertil.* **54,** 209–213.

Hoffman, R. A., and Robinson, P. F. (1966). Changes in some endocrine glands of white-tailed deer as affected by season, sex and age. *J. Mammal.* **47,** 266–280.

Hughes, E., and Mall, R. (1958). Relation of the adrenal cortex to condition of deer. *Calif. Fish Game* **44,** 191–196.

Jaczewski, Z., and Galka, B. (1967). Effects of administration of testosteronum propionicum on the antler cycle in red deer. *Finn. Game Res.* **30,** 303–308.

Jaczewski, Z., and Krzywinska, K. (1974). The induction of antler growth in a red deer male castrated before puberty by traumatization of the pedicle. *Bull. Acad. Pol. Sci., Ser. Sci. Biol.* **22,** 67–72.

Jaczewski, Z., and Michalakowa, W. (1974). Observations on the effect of human chorionic gonadotrophin on the antler cycle of fallow deer. *J. Exp. Zool.* **190,** 79–88.

Johnson, A. D., Gomes, W. R., and VanDemark, N. L. (1970). Early actions of cadmium in the rat and domestic fowl testis. I. Testis and body temperature changes caused by cadmium and zinc. *J. Reprod. Fertil.* **21,** 383–393.

Lambiase, J. T., Jr., Amann, R. P., and Lindzey, J. S. (1972). Aspects of reproductive physiology of male white-tailed deer. *J. Wildl. Manage.* **36,** 868–875.

Lebedinsky, N. G. (1939). Beschleunigung der Geweihmetamorphose beim Reh (Capreolus capreolus, L.) durch das Schilddrüsenhormon. *Acta Biol. Latv.* **9,** 125–132.

Lent, P. C. (1965). Observations on antler shedding by female barren-ground caribou. *Can. J. Zool.* **43,** 553–558.

Letellier, M. A., Plotka, E. D., Seal, U. S., Verme, L. J., and Ozoga, J. J. (1978). A technique for pinealectomy in deer, with notes on the neuroanatomy. *Am. J. Vet. Res.* **39,** 1617–1620.

Lincoln, G. A. (1971). The seasonal reproductive changes in the red deer stag (*Cervus elaphus*). *J. Zool.* **163,** 105–123.

Lincoln, G. A., and Kay, R. N. B. (1979). Effects of season on the secretion of LH and testosterone in intact and castrated red deer stags (*Cervus elaphus*). *J. Reprod. Fertil.* **55,** 75–80.

Lincoln, G. A., Guinness, F., and Short, R. V. (1971). The history of a hummel. Part 2. *Deer* **4,** 630–631.

Lincoln, G., Fletcher, J., and Guinness, F. (1976). History of a hummel. Part 4. The hummel dies. *Deer* **3,** 552–555.

McEwan, E. H., and Whitehead, P. E. (1972). Reproduction in female reindeer and caribou. *Can. J. Zool.* **50,** 43–46.

McMillin, J. M., Seal, U. S., Keenlyne, K. D., Erickson, A. W., and Jones, J. E. (1974). Annual testosterone rhythm in the adult white-tailed deer (Odocoileus virginianus borealis). *Endocrinology* **94,** 1034–1040

MacNally, L. (1976). Enigmas in antlers. *Deer* **3,** 476–481.

Magruder, N. D., French, C. E., McEwen, L. C., and Swift, R. W. (1957). Nutritional require-ments of white-tailed deer for growth and antler development. II. Experimental results of the third year. *Bull.—Pa., Agric. Exp. Stn.* **628,** 1–21.

Markwald, R. R., Davis, R. W., and Kainer, R. A. (1971). Histological and histochemical periodicity of cervine Leydig cells in relation to antler growth. *Gen. Comp. Endocrinol.* **16,** 268–280.

Millais, J. G. (1897). "British Deer and Their Horns." Henry Sotheran & Co., London.

Mirarchi, R. E., Howland, B. E., Scanlon, P. F., Kirkpatrick, R. L., and Sanford, L. M. (1978). Seasonal variation in plasma LH, FSH, prolactin, and testosterone concentration in adult male white-tailed deer. *Can. J. Zool.* **56,** 121–127.

Nicolls, K. E. (1971). A light microscopic study of nuclear and cytoplasmic size of the aggregate acidophil population in the hypophysis cerebri, pars distalis, of adult male mule deer, *Odocoileus hemionus hemionus*, relative to seasons of the photoperiod and antler cycles. *Z. Zellforsch. Mikrosk. Anat.* **115,** 314–326.

Pantić, V., and Stošić, N. (1966). Investigations of the thyroid of deer and roe-bucks. *Acta Anat.* **63,** 580–590.

Peterson, R. L. (1974). A review of the general life history of moose. *Nat. Can.* **101,** 9–21.

Pimlott, D. H. (1959). Reproduction and productivity of Newfoundland moose. *J. Wildl. Manage.* **23,** 381–401.

Plotka, E. D., Seal, U. S., Schmoller, G. C., Karns, P. D., and Keenlyne, K. D. (1977a). Reproductive steroids in the white-tailed deer (Odocoileus virginianus borealis). I. Seasonal changes in the female. *Biol. Reprod.* **16,** 340–343.

Plotka, E. D., Seal, U. S., Verme, L. J., and Ozoga, J. J. (1977b). Reproductive steroids in the white-tailed deer (Odocoileus virginianus borealis). II. Progesterone and estrogen levels in peripheral plasma during pregnancy. *Biol. Reprod.* **17,** 78–83.

Plotka, E. D., Seal, U. S., Letellier, M. A., Verme, L. J., and Ozoga, J. J. (1979). Endocrine and morphologic effects of pinealectomy in white-tailed deer. *In* "Animal Models for Research on Contraception and Fertility" (N. J. Alexander, ed.), pp. 452–466. Harper & Row, Hagerstown, Maryland.

Prior, R. (1968). "The Roe Deer of Cranborne Chase." Oxford Univ. Press, London/New York.

Ringberg, T. (1979). The Spitzbergen reindeer—a winter-dormant ungulate? *Acta Physiol. Scand.* **105,** 268–273.

Ringberg, T., Jacobson, E., Ryg, M., and Krog, J. (1978). Seasonal changes in levels of growth hormone, somatomedin and thyroxine in free-ranging, semi-domesticated Norwegian reindeer [Rangifer tarandus tarandus (L.)]. *Comp. Biochem. Physiol. A.* **60A,** 123–126.

Robinson, R. M., Thomas, J. W., and Marburger, R. G. (1965). The reproductive cycle of male white-tailed deer in central Texas. *J. Wildl. Manage.* **29,** 53–59.

Schulte, B. A., Parsons, J. A., Seal, U. S., Plotka, E. D., Verme, L. J., and Ozoga, J. J. (1980a). Heterologous radioimmunoassay for deer prolactin. *Gen. Comp. Endocrinol.* **40,** 59–68.

Schulte, B. A., Seal, U. S., Plotka, E. D., Verme, L. J., Ozoga, J. J., and Parsons, J. A. (1980b). Seasonal changes in prolactin and growth hormone cells in the hypophyses of white-tailed deer. *Am. J. Anat.* **159,** 369–377.

Schulte, B. A., Seal, U. S., Plotka, E. D., Letellier, M. A., Verme, L. J., Ozoga, J. J., and Parsons, J. A. (1981a). The effect of pinealectomy on seasonal changes in prolactin secretion in the white-tailed deer (Odocoileus virginianus borealis). *Enodcrinology* **108,** 173–178.

Schulte, B. A., Seal, U. S., Plotka, E. D., Verme, L. J., Ozoga, J. J., and Parsons, J. A. (1981b). Characterization of seasonal changes in prolactin and growth hormone cells in the hypophyses of white-tailed deer (Odocoileus virginianus borealis) by ultrastructural and immunocytochemical techniques. *Am. J. Anat.* **160,** 277–284.

Sempéré, A. (1977). Plasma progesterone levels in the roe deer, Capreolus capreolus. *J. Reprod. Fertil.* **50,** 365–366.

Sempéré, A. J., and Boissin, J. (1981). Relationship between antler development and plasma androgen concentrations in adult roe deer (Capreolus capreolus). *J. Reprod. Fertil.* **62,** 49–53.

Sempéré, A. J., and Boissin, J. (1983). Neuroendocrine and endocrine control of testicular activity and the antler cycle from birth to adulthood in the male roe deer (Capreolus capreolus L.). *In* "Antler Development in Cervidae" (R. D. Brown, ed.). Caesar Kleberg Wildl. Res. Inst., Kingsville, Texas (in press).

Sempéré, A., Garreau, J.-J., and Boissin, J. (1980). Variations saisonnières de l'activité de marquage territorial et de la testostéronémie chez le Chevreuil mâle adulte (Capreolus capreolus L.). *C. R. Hebd. Seances Acad. Sci.* **290,** 803–806.

Short, R. V. (1968). Factors controlling antler growth and development. *Deer* **1,** 218–221.

Short, R. V., and Hay, M. F. (1966). Delayed implantation in the roe deer Capreolus capreolus. *In* "Comparative Biology of Reproduction in Mammals" (I. W. Rowlands, ed.), pp. 173–194. Academic Press, New York.

Short, R. V., and Mann, T. (1966). The sexual cycle of a seasonally breeding mammal, the roebuck (*Capreolus capreolus*). *J. Reprod. Fertil.* **12,** 337–351.

Stieve, H. (1950). Anatomische Untersuchungen über die Fortpflanzungstätigkeit des europäischen Rehes (*Capreolus capreolus capreolus* L.). *Z. Mikrosk. Anat. Forsch.* **55,** 427–530.

Stošić, N., and Pantić, V. (1966). Cyclic changes in deer pituitary. *Yugosl. Physiol. Pharmacol. Acta* **2,** 231–237.

Suttie, J. M. (1980). Influence of nutrition on growth and sexual maturation of captive red deer stags. *Proc. Intn. Reindeer/Caribou Symp., 2nd, 1979* pp. 416–421.

Suttie, J. M. (1983). The influence of nutrition and photoperiod on the growth of antlers of young red deer. *In* "Antler Development in Cervidae" (R. D. Brown, ed.). Caesar Kleberg Wildl. Res. Inst., Kingsville, Texas (in press).

Tachezy, R. (1956). Über den Einfluss der Sexualhormone auf das Geweihwachstum der Cerviden. *Saugetierkd. Mitt.* **4**(3), 103–112.

Taylor, D. O. N., Thomas, J. W., and Marburger, R. G. (1964). Abnormal antler growth associated with hypogonadism in white-tailed deer in Texas. *Am. J. Vet. Res.* **25,** 179–185.

Thomas, D. C., and Cowan, I. McT. (1975). The pattern of reproduction in female Columbian black-tailed deer, *Odocoileus hemionus columbianus. J. Reprod. Fertil.* **44,** 261–272.

Topiński, P. (1975). Abnormal antler cycles in deer as a result of stress inducing factors. *Acta Theriol.* **20,** 267–279.

Ullrey, D. (1983). Nutrition and antler development in white-tailed deer. *In* "Antler Development in Cervidae" (R. D. Brown, ed.). Caesar Kleberg Wildl. Res. Inst., Kingsville, Texas (in press).

Ullrich, W. (1961). Zweimalige Geweihbildung in Jahresablauf bei einem Rehbock. *Zool. Garten, Leipzig* [N.S.] **25,** 411–412.

Verme, L. J. (1965). Reproduction studies on penned white-tailed deer. *J. Wildl. Manage.* **29,** 74–79.

West, N. O., and Nordan, H. C. (1976a). Hormonal regulation of reproduction and the antler cycle in the male Columbian black-tailed deer (*Odocoileus hemionus columbianus*). Part I. Seasonal changes in the histology of the reproductive organs, serum testosterone, sperm production, and the antler cycle. *Can. J. Zool.* **54,** 1617–1636.

West, N. O., and Nordan, H. C. (1976b). Cytology of the anterior pituitary at different times of the year in normal and methallibure-treated male Columbian black-tailed deer (*Odocoileus hemionus columbianus*). *Can. J. Zool.* **54,** 1969–1978.

Whitehead, P. E., and McEwan, E. H. (1973). Seasonal variation in the plasma testosterone concentration of reindeer and caribou. *Can. J. Zool.* **51,** 651–658.

Whitehead, P. E., and West, N. O. (1977). Metabolic clearance and production rates of testosterone at different times of the year in the male caribou and reindeer. *Can. J. Zool.* **55,** 1692–1697.

Wika, M. (1980). On growth of reindeer antlers. *Proc. Int. Reindeer/Caribou Symp., 2nd, 1979* pp. 416–421.

Wislocki, G. B. (1949). Seasonal changes in the testes, epididymes and seminal vesicles of deer investigated by histochemical methods. *Endocrinology* **44,** 167–189.

Castration

Castration is probably one of the oldest surgical procedures devised by man. It was particularly useful when wild animals became domesticated, because some of their more troublesome behavioral instincts associated with the mating seasons were thereby ameliorated. So it was that Aristotle wrote, "If stags be mutilated when, by reason of their age, they have as yet no horns, they never grow horns at all; if they be mutilated when they have horns, the horns remain unchanged in size, the animal does not lose them." Numerous experiments during the past century have confirmed what Aristotle already knew, that castration of fawns precludes antler growth altogether while castration of adults leads to the permanent retention of antlers.

Methods and Degrees of Emasculation

Castration is a relatively simple operation. One method, under suitable sedation, is to make a small incision in the scrotum, exteriorize the testes, ligate their stalks (including the major blood vessels) and cut them off distal to the ligation. Tying off the testicular arteries, in the absence of actual excision, results in testicular degeneration (Olt, 1927a,b). A convenient bloodless technique is "elastration" whereby a thick rubber band is applied around the neck of the scrotum to cause ischemic necrosis. The dead tissues drop off in a few weeks. In other species it has been shown that subcutaneous injections of cadmium chloride in appropriate concentrations specifically interferes with the blood flow to the testes, again resulting in their degeneration (Knorre, 1971). However, in such cases there may be subsequent regeneration of the Leydig cells from survivors immediately beneath the capsule. Although this may lead to recovery of the endocrine functions of the testes, there can be no regeneration of the seminiferous tubules. Presumably, such treatment of deer (not as yet tested) would have a transient influence on the antlers, but they would probably return to normal once the endocrine component of the testes had been reestablished.

It would be convenient for various reasons to be able to reverse the effects of castration, but this is obviously impossible after surgical removal of the testes. However, experiments on white-tailed deer with an antiandrogen, cyproterone acetate, have simulated this condition (Bubenik et al., 1975). If deer are treated with this compound, the production of testosterone is reduced or prevented, in turn precluding mineralization and shedding of the velvet which characterize the maturation of antlers. The effects are more pronounced if the drug is administered early in the antler growth period rather than during the later phases of development. Therefore, the effects of cyproterone acetate mimic the effects of castration, but are reversible when treatment is discontinued.

There are a number of recorded cases in which deer with abnormal antlers have been found to have atrophic testes. This condition has been reported in white-tailed deer by Taylor et al. (1964), mule deer by Robinette and Jones (1959), black-tailed deer by DeMartini and Connolly (1975), and moose by Lönnberg (1913). In these deer, the antlers were permanently in velvet, and strongly resembled those produced following castration. However, Marburger et al. (1967) described a white-tailed deer with atrophic testes and no evidence of spermatogenesis, but its antlers were bony. Because there was little or no scrotum, this may have been a case of cryptorchidism.

Cryptorchidism occurs when the testes fail to descend to the scrotal sac and are retained in the abdominal cavity. Here they are held at a higher temperature than when in the scrotal position. For reasons yet to be explained, normal spermatogenesis requires a temperature lower than that prevailing in the more central parts of the body. Therefore, cryptorchid testes possess degenerated seminiferous tubules, and, of course, are not fertile. However, their Leydig cells, which secrete testosterone, remain unaffected. As a result, a cryptorchid animal is hormonally normal but reproductively sterile. Normal antlers in cases of cryptorchidism have been reported in mule deer (Beauchamp and Jones, 1957) and reindeer (Leader-Williams, 1979). The condition has been studied experimentally in red deer by Zawadowsky (1926) who found, as expected, that antler growth was quite normal. Vasectomy would also be expected not to affect antler growth. There is some evidence (Lincoln, 1975) that if the epididymides are left intact after the testes themselves have been removed, the subsequently grown antlers will develop more branches. It is possible that the epididymis may be responsible for some testosterone secretion which could account for these effects.

There has been a persistent belief for many years that unilateral castration causes the development of an abnormal antler on the opposite side of the head (Clarke, 1916; Darling, 1937; Fowler, 1894; Zimmer, 1905). Such a "cross-effect" would be difficult to explain in terms of hormone influences, but the possibility of such a phenomenon cannot be ignored (Chapter 9). Examples of unequal antler production in unilaterally castrated deer have been cited occasionally in the literature, but there are other cases on record in which the loss of one testis has been shown not to affect the normal symmetry of the antlers. Indeed, specific experiments designed to test this possibility have effectively disproven the older claims (Jaczewski, 1952; Penrose, 1924; Zawadowsky, 1926).

The Castrate Antler

The precise age at which castration ceases to prevent antler growth and allows the production of permanent ones has not been determined. It undoubtedly varies with the species. In the red deer, there is evidence that castration any time prior to puberty will preclude antler growth (Jaczewski et al., 1976; Zawadowsky, 1926). In the fallow deer, castration during the first few weeks of life usually prevents antlers from growing (Zawadowsky, 1926), but in at least one case such an animal castrated at birth is reported to have grown "simple dags" (Fowler, 1894). Castration at the age of several months may allow the development of complete, but permanent, antlers (Zawadowsky, 1926). In the roe deer, a species which sometimes grows its initial, but very short, set of antlers as a fawn, castration soon after birth interferes with such outgrowths (Blauel, 1935, 1936). If the operation is delayed as little as 1 month, abnormal antlers may be produced (Rörig, 1907). Reindeer and caribou are unique in that they begin to produce their first sets of antlers within several weeks of being born. Castration of these calves nevertheless permits the growth of antlers, but they are small and abnormal (Luick, 1978).

If castration of adults is conducted in the autumn or winter when the antlers are bony, the first effect is the premature casting of the antlers. These antlers can be distinguished by the flat or concave base where they were detached from the pedicle, as opposed to the convex base typical of normally cast antlers (Fig. 142). Ordinarily, the old antlers are not lost until winter or spring, depending on the species. This event is correlated with the seasonal decrease in testosterone production after rut. Castration simply accelerates what happens naturally by bringing about an abrupt and premature decline in testosterone levels. The earlier in the season castration is performed, the longer it takes for the old antlers to drop off. About 1 month may be required early in the fall, while casting may occur 1–2 weeks after castration in the early spring. The interval is probably a function of the length of time it takes for the residual testosterone to be cleared from the system and the time required for the osteoclastic mechanism to react.

One of the basic unanswered questions of antler growth is whether or not there is a causal relation between casting of the old antlers and the onset of new growth. In the case of castration, renewed antler growth occurs as soon as the old antlers have been lost, even in those species in which there is normally a lag between these two events. Reindeer, caribou, moose, and members of the genus *Odocoileus* all normally exhibit a delay of up to several months between casting and regrowth. These exceptional cases do not disprove the possibility that a single hormone, or absence of same, might be responsible for both events (perhaps by differential reactions to changing levels in the blood stream), but neither are they inconsistent with the possibility that casting and regrowth might be attributed to two different hormones. This possibility is strengthened by certain reactions exhibited by the antlers of castrated deer.

The antlers grown by castrated animals are deceptively normal in appearance the first time around. They may be a little thinner than normal, with fewer branches, and not infrequently their angles of growth are un-

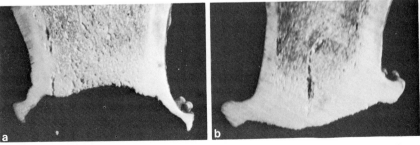

Fig. 142. Comparison of the bases of cast antlers, viewed in longitudinal section, from castrated (a) and intact (b) deer. (From Goss, 1963. Copyright 1963 by the American Association for the Advancement of Science.)

usual. Those in castrated reindeer are said to be larger and heavier than in intact animals (Tandler, 1910). Otherwise, all seems normal except that they fail to undergo the final stages of maturation. Ossification is limited and the velvet is not shed. Therefore, the antlers of castrated deer are composed of bone that is more porous than the solid consistency of normal mature antlers.

What happens next depends on environmental conditions. Deer living in temperate climates are vulnerable to the freezing effects of winter weather. The antlers of castrated deer are especially sensitive to this, and they often freeze down to the last few centimeters above the burr. In milder conditions, substantially greater lengths of the antler may survive freezing. In either case, the frozen ends of the antlers become necrotic and are lost in the spring (Fig. 143). The mechanism by which the dead portions of the antler are cast is superficially reminiscent of that responsible for the normal loss of bony antlers, although the cellular mechanisms of these processes have not been compared. Wound healing occurs on the exposed stumps and re- newed antler growth takes place, but not always in the usual way.

A deer which is stimulated to produce new antler tissue without having lost all of its old antlers is destined to grow some rather bizarre structures. The actual shapes of these castrate antlers vary with the species and the climate. In general, there is a strong tendency for amorphous outgrowths to be produced around the base of the antler (Fig. 144), but tuberosities may also be grown on the sides of the shafts. These outgrowths, often relatively modest at first, may grow to exaggerated dimensions in subsequent years. The extent of their growth is enhanced in deer whose antlers are prevented from freezing by being kept indoors during the winter months. Under these circumstances each year's growths are produced on top of previous incre- ments, often leading to grotesque, tumorlike masses extending down over the head of the animal. Similar unsightly growths may sprout from the ends of the antlers, particularly on the lateral sides.

Although castrate antlers have not been described in all species of deer, there is reason to believe that similar reactions occur throughout the Cer- vidae. Even in the axis deer, a tropical species, typical episodes of growth have been described by Bullier (1948) in a castrated specimen in France. Each winter the ends of the velvet antlers froze, to be replaced by new outgrowths in the summer. Caton (1877) and Skinner (1923) recorded the reactions of wapiti following castration. Although their antlers tend to be too large to freeze completely, they may lose their points which usually heal over without regenerating. Eventually, these antlers may double their nor- mal diameters, lose their branches, and grow tuberosities on their surfaces, especially from the burr.

Castrated mule deer in southwestern United States have been referred to

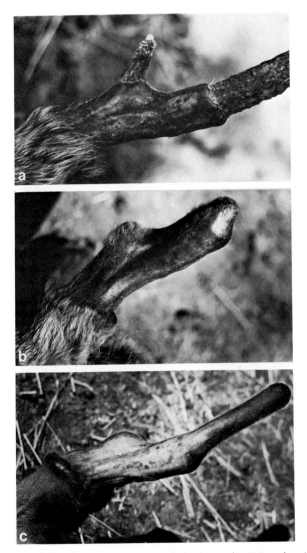

Fig. 143. Necrosis and regrowth of the castrated sika deer antler. Winter freezing has killed the distal portions of the velvet antler (March 13) (a) which have dropped off and healed by April 26 (b). On June 7 the missing antler is regenerating (c).

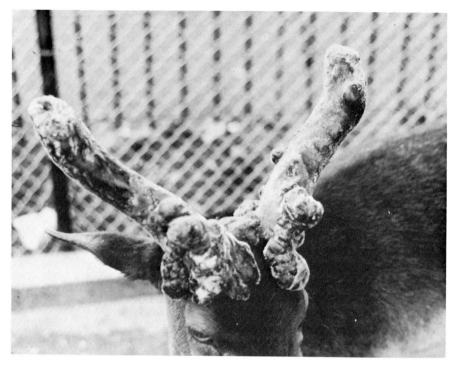

Fig. 144. Overgrowth of antlers in a castrated fallow deer formerly residing in Regent's Park Zoo, London. Relatively mild winters allowed survival of much of the main beams, the widths of which increased yearly. Basal tumors have grown down over the head.

as "cactus bucks" because of the swollen appearance of their antlers (Mearns, 1907). A white-tailed deer in New York developed stunted antlers after castration (Townsend and Smith, 1933), primarily because any new growths tended to be frostbitten each winter. However, a specimen from Maine grew a cluster of amorphous masses that capped his head (Fig. 145). Waldo and Wislocki (1951) and Wislocki et al. (1947) described a castrated white-tailed deer held indoors each winter at the Philadelphia zoo for 9 years. This animal sported a profusion of branches radiating from extra large pedicles on its head (Fig. 146). Such luxuriant growths could not have survived winter freezings, but they shed their velvet after testosterone treatment and were later detached and replaced by equally abnormal outgrowths. It is curious that castrate antlers sometimes sprout extra branches or regenerate missing parts (Figs. 143 and 146), but in other instances they just grow thicker and perhaps develop tumorous swellings (Figs. 144 and 145). Controlled investigations are needed to determine how these differing responses are to be explained.

Fig. 145. White-tailed deer shot near Bucksport, Maine. Castration has been responsible for the growth of tuberosities in lieu of normal antlers. (Courtesy of Dr. Charles Mack.)

Fig. 146. Castrated white-tailed deer protected from winter freezing for 9 years. Numerous extra points have developed instead of amorphous outgrowths. (After Wislocki *et al.,* 1947.)

Although the foregoing reactions are encountered in most species of castrated deer, the overgrowths of roe deer antlers following castration are in a class by themselves (Blauel, 1935, 1936; Kleesiek, 1953; Krieg, 1946; Olt, 1927a,b; Rörig, 1907; Zimmer, 1905). In this species, the extraordinary extent to which antler tissue grows following castration yields a mass of tissue which may seriously incapacitate the deer. Because it tends to cover the top of the head like a wig, these formations are often referred to as "perukes." In extreme cases, outgrowths may hang down in front of the eyes, interfering wth the animal's vision (Fig. 147). Even worse, these peruke antlers are vulnerable to infections and infestations (Bickel, 1936). Necrotic tissues are eaten out by maggots in some cases (Rörig, 1907). Each year the peruke antler increases in size, sometimes to five times the weight of the head (Olt, 1927a). The roe deer cannot long survive under these conditions.

One wonders why the reactions of roe deer antlers are so much greater than in other species. It may be related to the fact that roe deer are unique in that they normally grow their antlers in the winter instead of the warmer

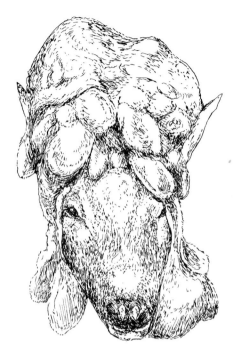

Fig. 147. Peruke antlers of a castrated roe deer. (After Tandler and Grosz, 1913.)

months of the year. Accordingly, their castrate antlers are not subject to freezing to the same extent as in other species. A more rapid accretion of antler tissue would be expected under such circumstances. Alternatively, the exuberant growth of castrate roe antlers could be attributed to the unusually prominant pearlation that normally adorns the shafts in this species, a condition that might predispose the roe deer to peruke formation.

Finally, it is worth exploring the reactions of reindeer and caribou to castration. For centuries the Lapps have castrated their reindeer bulls, although their methods have been at best primitive. According to Sjenneberg and Slagsvold (1968),

> The method consists of having one person with good teeth bite the testicles crosswise. The testicle tissue is kneaded afterward with the hand into a porridgelike consistency. Thus, the operation is performed without bloodshed and there is no danger of infection. Undoubtedly this method involves extensive suffering by the animal which tends to wander aimlessly for a couple of weeks after such a treatment.

Not surprisingly, it is difficult to interpret the results of such incomplete castrations (Hadwen and Palmer, 1922; Jacobi, 1931). However, under more controlled circumstances it is known that when reindeer are castrated while their antlers are bony, casting occurs after the usual several weeks' interval (Hadwen and Palmer, 1922; Sjenneberg and Slagsvold, 1968). However, it is not clear from records thus far available whether or not such deer grow new sets of antlers right away as has been shown to be the case in other deer.

When reindeer are castrated while in velvet, their antlers are retained through the fall and winter. However, unlike other species, these antlers do not become overgrown. They may be larger and heavier than normal (Tandler, 1910), and stay in velvet throughout the winter months (Bubenik et al., 1976), but the skin is in a desiccated condition (Meschaks and Nordkvist, 1962). These antlers are cast in the spring, later than in the case of intact males (Fisher, 1939), whereupon new antler growth is initiated. Therefore, in reindeer and caribou the antlers are renewed annually despite castration. The same is true of females from which the ovaries have been removed (Jacobi, 1931).

It is not known whether or not the velvet antlers remain viable through the arctic winter, but this would seem unlikely. Thus, the loss of antlers in the spring may be more akin to the passive separation of necrotic portions of castrate antlers in other species than to the active casting of bony antlers by normal deer. In any case, the occurrence of overgrowth of reindeer and caribou antlers following castration is not known. However, if such animals were to be protected from winter freezing, their antlers should not only remain perennially viable but might even give rise to perukelike outgrowths.

The only "cure" for castrate antlers is to treat the animal with testosterone or estrogen, each of which appears to be equally effective (Blauel, 1935; Blankenburg and Stocksmeier, 1958; Fletcher and Short, 1974; Kleesiek, 1953; Lincoln, 1975; Wislocki et al., 1947). These hormones induce belated ossification of the shaft and tines, followed by shedding of the velvet and associated overgrowths within a few weeks. Then, when the hormone wears off, the antlers themselves are cast. However, new growth leads eventually to equally abnormal antlers in future episodes.

The Antleroma: A Model for Tumor Growth?

The nature of the amorphous growths produced by the antlers of castrated deer is difficult to interpret. Because of their tumorlike appearance, they may appropriately be referred to as "antleromas" (Fig. 148). If they were true neoplasms, they would probably be classified as sarcomas. Yet there is no indication that they metastasize to other parts of the body (Olt, 1927a). In

Fig. 148. Photograph of a castrated fallow deer approximately 4 years after operation, by which time the antleromas have grown to impressive proportions, including a pendulous one that dangled in front of the eye tethered by a long stalk of skin.

the absence of evidence to the contrary, they are most logically considered benign tumors.

Studies on the histogenesis and differentiation potential of antleromas are long overdue. Because of the location of these growths on the surface of the antler, they would appear to be derived from either the dermis of the velvet or the underlying periosteum. Deletion and transplantation experiments would be useful in making this distinction. However, the failure of preliminary investigations to reveal either cartilage or bone in antleromas (Fig. 149) would argue against the possible periosteal origin of these tissues. This does not necessarily rule out this alternative, especially if the potential for chondrogenesis or osteogenesis were latent in castrated deer. For example, if skeletal differentiation of antleroma tissues were dependent on testosterone, this could not be expressed because testosterone brings about the death of the entire antler, including its dependent offshoots. Only by grafting antleromas, or portions thereof, to other parts of the body might they be protected from the testosterone-induced ischemic necrosis of the parent antler. The possible induction of differentiation by this hormone could then be studied in antleroma tissues allowed to survive long enough to react.

It is equally plausible that antleromas are derived from the dermis. The profusion of fibroblastlike cells and the quantities of interstitial collagen in these tumors is not inconsistent with their possible dermatogenic origin. If

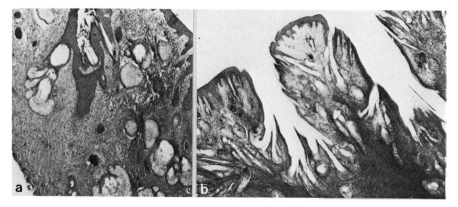

Fig. 149. Histology of antleroma. Undifferentiated cells populate the inner substance of these overgrowths (a). They are enveloped in velvet skin sometimes thrown into prominent papillae (b). Hematoxylin and eosin. (Photos by Nancy Bell.)

the dermal histogenesis theory were correct, then the possible identity of antleromas with keloids or hypertrophic scars would be an obvious interpretation of their basic nature.

The question then remains as to what stimuli trigger the onset of antleroma growth. Castration per se creates the endocrine conditions conducive to the development of antler tumors, but is presumably not responsible for the induction of outgrowths at specific locations. The region around the basal burr is especially predisposed to antleroma formation, as are the lateral sides of the antler beam. However, all surfaces of the antler are responsive, especially in the formation of peruke antlers in the castrated roe deer. In the fallow deer, incipient nodules may be observed crowding virtually all exposed surfaces of the velvet. Some of these mushroom into large tumors; others remain in abeyance. Whether vascular or neural influences determine the sites where these tuberosities originate is as yet unexplored. Nor has the possible role of mechanical trauma been investigated as a factor in promoting antleroma induction. Clearly, this intriguing phenomenon is ripe for experimental exploitation.

It is significant that antleromas, like the antlers from which they are derived, do not grow continuously. They are responsive to seasonal photoperiodic changes, ceasing growth in the late summer and resuming it in spring. If testosterone levels were nil in castrated deer, one would be inclined to attribute the cycles of antleroma growth to other sources of hormones, such as the pituitary, hypothalamus, or pineal glands. However, radioimmunologic assays of plasma from castrated white-tailed deer (Bubenik, 1983) have shown the persistence of low titres of testosterone, perhaps of adrenal cortical origin. If the levels of this hormone were to fluctuate seasonally, it could possibly explain the intermittent growth of antleroma tissue even in castrated males.

Antlers and Cancer

The reaction of antlers to castration has little parallel elsewhere in living systems, except insofar as antleromas may resemble cancer. Like the uncontrolled growth of tumors, the antlers of castrated deer grow in an almost unrestricted manner. They are an example of a tissue whose growth is regulated more by inhibitors than stimulators. In the absence of inhibition by sex hormones, they undergo an uncontrolled expansion which leads to the production of tumorous masses with little or no morphogenesis. Castrate antlers are a pathological condition, but except for the secondary consequences of infection and necrosis, and a certian energy drain on the system, they are not necessarily life threatening. Despite their superficial resemblance to cancer, they exhibit none of the malignant characteristics of invasiveness or metastasis that one associates with cancer (Gruber, 1937; Olt, 1927a). Although they would appear to be an example of unlimited growth in the absence of natural inhibitors, they are in fact not entirely without constraints. The seasonal cyclicity which characterizes their growth pattern implies the operation of stimulatory influences. Therefore, the well-known inhibitory effect of sex hormones on antler growth may be balanced by a hitherto undefined stimulatory one. Clearly there remains much to be learned regarding the nature of antler growth. The relatively neglected phenomenon of castrate antlers is a useful model system in which to study important developmental pathologies.

References

Aristotle (1910). "Historia Animalum" (Engl. transl. by D'A. W. Thompson), p. 631. Oxford Univ. Press (Clarendon), London/New York.

Beauchamp, D. E., and Jones, F. L. (1957). A cryptochid Rocky Mountain mule deer. *J. Mammal.* **38**, 423.

Bickel, E. (1936). Drei Perückenböcke. *Wild Hund* **42**, 674–675.

Blankenburg, W., and Stocksmeier, H. (1958). Hormonbehandlung bei einem Perückenbock. *Z. Jagdwiss.* **4**, 99–101.

Blauel, G. (1935). Beobachtungen über die Enstehung der Perücke beim Rehbock. *Endokrinologie* **15**, 321–329.

Blauel, G. (1936). Beobachtungen über die Entstehung der Perücke beim Rehbock. 2. Mitteilung. *Endokrinologie* **17**, 369–372.

282   13.   Castration

Bubenik, A., Tachezy, R., and Bubenik, G. (1976). The role of the the pituitary-adrenal axis in the regulation of antler growth processes. *Saugetierkd. Mitt.* **40**, 1–5.

Bubenik, G. A. (1983). Endocrine regulation of the antler cycle. *In* "Antler Development in Cervidae" (R. D. Brown, ed.). Caesar Kleberg Wildl. Res. Inst., Kingsville, Texas (in press).

Bubenik, G. A., Bubenik, A. B., Brown, G. M., and Wilson, D. A. (1975). The role of sex hormones in the growth of antler bone tissue. 1. Endocrine and metabolic effects of antiandrogen therapy. *J. Exp. Zool.* **194**, 349–358.

Bullier, P. (1948). Curieuse anomalie des bois chez un cerf castré (*Axis axis*). *Mammalia* **22**, 271–274.

Caton, J. D. (1877). "The Antelope and Deer of America." Hurd & Houghton, Cambridge, Massachusetts.

Clarke, F. C. (1916). Malformed antlers of deer. *Calif. Fish Game* **2**, 119–123.

Darling, F. F. (1937). "A Herd of Red Deer. A Study in Animal Behaviour." Oxford Univ. Press, London/New York.

DeMartini, J. C., and Connolly, G. E. (1975). Testicular atrophy in Columbian black-tailed deer in California. *J. Wildl. Dis.* **11**, 101–106.

Fisher, C. (1939). The nomads of arctic Lapland. *Natl. Geogr. Mag.* **76**, **641–676.**

Fletcher, T. J., and Short, R. V. (1974). Restoration of libido in castrated red deer stag (*Cervus elaphus*) with oestradiol-17β. *Nature (London)* **248**, 616–618.

Fowler, G. H. (1894). Notes on some specimens of antlers of the fallow deer, showing continuous variation, and the effects of total or partial castration. *Proc. Zool. Soc. London* pp. 485–494

Goss, R. J. (1963). The deciduous nature of deer antlers. *In* "Mechanisms of Hard Tissue Destruction" (R. F. Sognnaes, ed.), Publ. No. 75, pp. 339–369. Am. Assoc. Adv. Sci., Washington, D.C.

Gruber, G. B. (1937). Morphobiologische Untersuchungen am Cerviden-Geweih. Werden, Wechsel und Wesen des Rehgehörns. *Nachr. Ges. Wissensch. Goettingen, Math.-Phys. Kl., Fachgruppe 6* [N.S.] **3**(2), 9–63.

Hadwen, S., and Palmer, L. J. (1922). Reindeer in Alaska *U.S., Dep. Agric., Bull.* **1089.**

Jacobi, A. (1931). Das Rentier. *Zool. Anz.* **96**, 1–264.

Jaczewski, Z. (1952). Effect of unilateral castration on antler growth in roe-deer (*Capreolus capreolus* L.) and common stag (*Cervus elaphus* L.). *Fragm. Faun. Mus. Zool. Pol.* **6**, 199–205.

Jaczewski, Z., Doboszyńska, T., and Krzywiński, A. (1976). The induction of antler growth by amputation of the pedicle in red deer (*Cervus elaphus* L.) males castrated before puberty. *Folia Biol. (Krakow)* **24**, 299–307.

Kleesiek, C. (1953). Hormonbehandlung eines Perückenbocks. *Wild Hund* **56**, 117.

Knorre, D. (1971). Induction of interstitial cell tumors by cadmium chloride in albino rats. *Arch. Geschwulstforsch.* **38**, 257–263.

Krieg, H. (1946). Das Hirschgeweih. *Naturwissenschaften* **33**, 175–180.

Leader-Williams, N. (1979). Abnormal testes in reindeer, *Rangifer tarandus. J. Reprod. Fertil.* **57**, 127–130.

Lincoln, G. A. (1975). An effect of the epididymis on the growth of antlers of castrated red deer. *J. Reprod. Fertil.* **42**, 159–161.

Lönnberg, E. (1913). On a hypospadic pseudohermaphroditic elk. *Ark. Zool.* **7**(34), 1–8.

Luick, J. R. (1978). Why castrate? *Reindeer Herders Newsl.* **3**(2), 1–6.

Marburger, R. G., Robinson, R. M., and Thomas, J. W. (1967). Genital hypoplasia of white-tailed deer. *J. Mammal.* **48**, 674–676.

Mearns, E. A. (1907). Mammals of the Mexican boundary of the United States. *Bull.—U. S. Natl. Mus.* **56.**

Meschaks, P., and Nordkvist, M. (1962). On the sexual cycle in the reindeer male. *Acta Vet. Scand.* **3,** 151–162.

Olt, A. (1927a). Die Perücke des Cervidengeweihes und ihre Bedeutung für die Krebsforschung. *Ber. Oberhess. Ges. Nat.-Heilkd. Giessen, Naturwiss. Abt.* [N.S.] **11,** 3–7.

Olt, A. (1927b). Die Perücke der Cerviden und das Karzinom. *Tierärztl. Wochenschr.* **35,** 131–133.

Penrose, C. B. (1924). Removal of the testicle in a sika deer followed by deformity of the antler on the opposite side. *J. Mammal.* **5,** 116–118.

Robinette, W. L., and Jones, D. A. (1959). Antler anomalies of mule deer. *J. Mammal.* **40,** 96–108.

Rörig, A. (1907). Gestaltende Correlationen zwischen abnormer Körperkonstitution der Cerviden und Geweihbildung derselben. *Arch. Enicklungs Mech. Org.* **23,** 1–150.

Sjenneberg, S., and Slagsvold, L. (1968). "Reindeer Husbandry and Its Ecological Principles." (Engl. transl.) (C. M. Anderson and J. R. Luick, eds.). U. S. Dept. Interior, Bureau of Indian Affairs, Juneau, 1979.

Skinner, M. P. (1923). A castrated elk. *J. Mammal.* **4,** 252.

Tandler, J. (1910). Bericht . . . über den Einfluss der Geschlechtsdrüsen auf die Geweihbildung bei Rentieren vor. *Anz. Akad. Wiss. Wien, Math-Naturwiss. Kl.* **47,** 252–257.

Tandler, J., and Grosz, S. (1913). "Die Biologischen Grundlagen der Sekundären Geschlechtscharaktere." Springer-Verlag, Berlin/New York.

Taylor, D. O. N., Thomas, J. W., and Marburger, R. G. (1964). Abnormal antler growth associated with hypogonadism in white-tailed deer in Texas. *Am. J. Vet. Res.* **25,** 179–185.

Townsend, M. T., and Smith, M. W. (1933). The white-tailed deer of the Adirondacks. *Roosevelt Wild Life Bull.* **6,** 161–325.

Waldo, C. M., and Wislocki, G. B. (1951). Observations on the shedding of the antlers of Virginia deer (*Odocoileus virginianus borealis*). *Am. J. Anat.* **88,** 351–396.

Wislocki, G. B., Aub, J. C., and Waldo, C. M. (1947). The effects of gonadectomy and the administration of testosterone propionate on the growth of antlers in male and female deer. *Endocrinology* **40,** 202–224.

Zawadowsky, M. M. (1926). Bilateral and unilateral castration in *Cervus dama* and *Cervus elaphus*. *Trans. Lab. Exp. Biol. Zoo Park Moscow* **1,** 18–43.

Zimmer, A. (1905). Die Entwicklung und Ausbildung des Rehgehörns, die Grösse und das Körpergewicht der Rehe. *Zool. Jahrb., Abt. Syst. (Oekol.), Geogr. Biol.* **22,** 1–58.

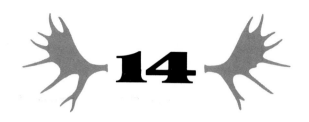

Antlered Does and
Antlerless Bucks

It has been said that nature's exceptions are the germs of discovery. If so, the study of those exceptional instances in which female deer grow antlers, or in which males do not, may yield clues to the fundamental nature of antler growth in general. The fact that female deer do on rare occasions grow antlers spontaneously is an intriguing phenomenon that demands explanation. The eventual solution to the problem may come from two approaches. One is the careful analysis of those cases that occur naturally. The other is the duplication of such freaks by experimentation on captive deer.

Reindeer and Caribou

It is important to note that female reindeer and caribou normally carry antlers. Nevertheless, even in these deer there may be a significant incidence of females without them. In Newfoundland, for example, 55% of the female woodland caribou in the interior lack antlers, while on the Avalon peninsula 91% may be antlerless (Bergerud, 1976). Such animals have also been reported by Murie (1935) in Alaska. In all probability, this condition is genetically determined, but the selective pressures responsible for the presence or absence of antlers in reindeer and caribou are not easily identified. It may be significant that barren ground caribou, as well as reindeer, seldom if ever lack antlers (Seton, 1929).

The explanation may lie in the social structure of the population. More than any other species of deer, reindeer and caribou live in herds, a characteristic shared with many species of horned ungulates. In the latter animals the presence of horns in females, albeit smaller than those of the males, appears to be correlated with the herding instinct (Chapter 5). More solitary species tend to restrict horns to the males. One wonders if the importance of horns and antlers in females of the species is to provide competitive advantage with males in the give and take of herd life. Such advantages are enhanced by the retention of antlers in female reindeer and caribou until their calves are born in the spring, whereas the males lose their antlers in winter. Such an arrangement could be interpreted to enhance the social dominance of pregnant cows during those important last months of pregnancy. The fact that barren females drop their antlers earlier than pregnant ones is consistent with this interpretation.

Incidence of Female Antlers

Antlered does have been reported in four genera. They are relatively common in roe deer and in white-tailed and mule deer. They are much rarer in sika, wapiti, red deer, and moose. During the 4-year period from 1957 to 1960 in Pennsylvania, 39 antlered "does" were shot (by mistake) out of 173,038 white-tailed bucks killed by hunters during that period (Doutt and Donaldson, 1961). If one assumes an even sex ratio in the population, then about 1 of every 4400 does in Pennsylvania carries antlers. Ryel (1963) reported incidences among white-tailed deer of up to 1 antlered doe per about 900 adult females in Michigan. In this and other species in which the phenomenon has been observed, such antlers are almost always unbranched spikes, although bifurcation is observed on rare occasions (Berry, 1932; Buss, 1959; Dixon, 1927). Sometimes such does will carry an antler only on one side of her head (Buss and Solf, 1959), and when the occurrence is bilateral the two antlers are commonly different lengths (Whitehead, 1971). Female antlers are usually short, but lengths of 13 cm have been reported in the mule deer, and 29 cm in the white-tail (Doutt and Donaldson, 1959b; Haugen and Mustard, 1960). One female roe deer possessed antlers 15 cm long (Alson, 1879), and a wapiti doe had a single 47-cm antler on her left side (Buss and Solf, 1959). A female sika deer grew antlers 10 and 15 cm long (Roux and Stott, 1948).

The majority of antlered does are encountered in the late autumn when they are shot during the hunting season (Fig. 150). This is the season when bucks normally possess burnished antlers, having shed their velvet in late summer. Nevertheless, most of the antlered does observed at this time of year carry antlers that are in velvet (Donaldson and Doutt, 1964; Doutt and Donaldson, 1959a, 1960; Mierau, 1972; Wislocki, 1954, 1956). This may be taken to indicate that the hormonal conditions that are normally responsi-

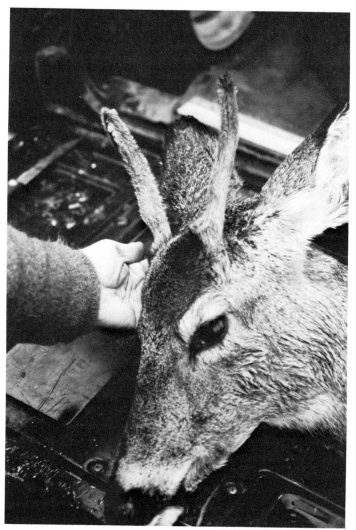

Fig. 150.   An antlered white-tailed doe with unbranched antlers in velvet when she was shot in December. Her reproductive tract was normal.

ble for the maturation of the antler by inducing ossification and shedding of the velvet, are not functioning properly in females. However, in the mule deer there have been several cases in which females carried polished antlers in the early autumn.

There is as yet no consensus on whether or not female antlers are cast and replaced each year. The usual failure of such antlers to complete their maturation by shedding the velvet would suggest that these antlers, like those of castrated bucks, might be retained indefinitely (Diem, 1958). Changes in the length of such antlers from year to year could be explained by the loss of their ends as a result of freezing in the winter, possibly followed by regrowth the next year. Although the presence of the basal burr in female antlers has been cited by Mierau (1972) as proof that casting and regrowth have occurred, this structure is sometimes developed on an animal's first set of antlers anyway. Therefore, it would seem prudent to assume that the spontaneously grown antlers of female deer (except for reindeer and caribou) are retained from one year to the next pending firsthand observations to the contrary.

The Fertility of Antlered Does

The presence of antlers in female deer does not apparently interfere with their reproductive potentials (Doutt and Donaldson, 1959a; Mierau, 1972). The vast majority of does with antlers still in velvet in the autumn showed unmistakable evidence of having given birth to fawns the previous spring. They were either accompanied by their offspring (Diem, 1958; Haugen and Mustard, 1960), or showed signs of having lactated (Buechner, 1957; Doutt and Donaldson, 1960). Those taken later in the season had recently ovulated or were already pregnant. However, occasional does with mature, bony antlers show no evidnece of having reproduced (Doutt and Donaldson, 1961).

Thus, the question is what is the cause of antler production in does. It could either be a genetically inherited trait (as is undoubtedly the case in reindeer and caribou) or the result of a pathological condition. In an effort to explore the latter possibility, Doutt and Donaldson (1959a) surveyed a large number of antlered does shot in the 1950s by Pennsylvania deer hunters during the December hunting season. In most instances the females were otherwise normal. However, in rare cases of antlered does, abnormalities of the reproductive tract would appear to have been the cause. At least one "doe" was found to be a hermaphrodite, an animal possessing both ovary and testis (Donaldson and Doutt, 1964; Doutt and Donaldson, 1961). Wislocki (1956) described the case of a deer with antlers that had shed their velvet, and which from external appearances was a female. Internally, however, the uterus was not fully developed and there was an undescended testis on one side. In animals with such disordered sex organs, there may be just enough of a male influence to promote the development of pedicles in the fawns, followed by antler growth. Finally, there has been one case of a female bearing a tumor in her adrenal cortex (Doutt and Donaldson, 1959b). Since comparable tumors have been shown to exert masculinizing effects in women, it is not surprising that such a tumor in a deer might secrete enough testosterone to be responsible for the induction of antlers.

The presence of antlers in female deer is a reminder of how fragile the distinction between the sexes can be. Although the gender of an individual is genetically determined, the extent to which sexual characteristics are in fact expressed may be affected by a number of physiological conditions. One indication of the genetic sex is the presence of the sex chromatin. This can be seen under the microscope as a darkly stained body next to the nucleus of the cell. Sex chromatin is derived from one of the X chromosomes, and is diagnostic of female cells. An antlered doe investigated for this character was found to possess sex chromatin in her cells (Crispens and Doutt, 1973).

Despite the genetic sex of these individuals, the expression of this characteristic could conceivably be modified even before birth. This is what happens in the case of the freemartin. A freemartin is a heifer born cotwin to a bull. The condition has been known since the middle ages, and the name refers to the fact that its flesh is good only for the market, not for breeding. Such animals are genetic females, but the development of their reproductive tracts has been abnormally affected by male influences. These influences derive from their brothers in the womb. Sometimes the placenta of such twins become fused, as a result of which their bloodstreams intermingle. When the male twin begins to develop its reproductive tract, the hormones produced can circulate into his sister's bloodstream and exert effects on the development of her genitalia. As a result of this, she may become a partially masculinized female and therefore infertile. Although the freemartin is typical of cattle, prenatal human blood chimeras have been reported in medical annals. Therefore, it is possible that antlered does might have been subjected to the prenatal influences of male hormones if they had had twin brothers and their placentas had been fused in the uterus. There is no proof that such occurrences take place, but it is perhaps significant that the incidence of antlered does seems to be roughly correlated with the extent to which different species of deer tend to give birth to twins. Multiple births are common in white-tailed and mule deer, and it is in these species that the largest number of antlered does has been reported. Twinning also occurs in roe deer, the females of which sometimes grow antlers. However, in the moose twins are not unusual but antlers have not as yet been reported in females of this species of deer. The production of antlers by females of the genus *Cervus,* such as the sika, red deer and wapiti, is a relatively rare phenomenon, matched only by the infrequency with which these species give birth to twins. Notwithstanding the foregoing correlations, it must be emphasized that there is no direct evidence that antlered females are in fact the cervid counterparts of freemartins. As reviewed by Wurster and Benirschke (1967), placental fusion may occur in twin fawns, but vascular anastomoses have been reported only in the moose (Kurnosov, 1962), not in the roe deer (Hamilton *et al.,* 1960) or white-tailed deer (Wurster and Benirschke 1967). The otherwise normal state of the reproductive organs in almost all cases of antlered does would suggest that the consequences of twinning are probably not the true explanation for the masculinization of female deer.

The major obstacle to the resolution of the problem of antlered does has been the fact that this phenomenon is difficult to study in captive animals. It comes to light only after the deer in question has been killed. However, the logical approach would be to study a live specimen, preferably through more than one generation. Such an opportunity was reported by Mierau (1972) who discovered an antlered female mule deer in Colorado in 1968. Because she was within Mesa Verde National Park, she was protected from being shot by hunters. On May 7, 1968, she possessed a 10-cm-long unbranched spike in velvet on her right side. She was accompanied by a female yearling, presumably her daughter from the year before, and she had twins for the next 3 years. When observed on September 30, 1968, she possessed antlers on both sides, the right one 13 cm long and the left one 8 cm. In 1970, the same deer possessed antlers slightly more than 4 cm in length. Although these antlers had burrs at their bases, it is not known that they had actually replaced previous ones since the ends of her original antlers might have been lost as a result of freezing. She was shot in June of 1970, at the age of 5–7 years, and was found to have normal visceral organs. A sample of her blood cells was grown in culture to study the chromosomes. This showed a normal karyotype of 70 chromosomes, including the female complement of two X chromosomes. In 1963, Benirschke *et al.* had likewise failed to find chromosomal abnormalities in two white-tailed does with velvet antlers. These findings would rule out the possibility that antlered does might have inherited a Y chromosome by mistake from the father in addition to their normal pair of X chromosomes. The most significant aspects of the case of the Mesa Verde doe is that her female offspring could be observed, and one of these was found also to have grown a set of antlers 5 cm long that were in velvet in December, 1968. This strongly suggests that, despite the normal appearance of the chromosomes, antler production by female deer is probably a genetic trait passed on from one generation to the next. This conclusion is supported by the fact that in the same area of southern Rhode Island, a state not known for its large deer population, antlered does were shot by deer hunters in both 1976 and 1977. One wonders if these two white-tailed females, which were in velvet in December and were reproductively normal, might not have been related to each other. A similar situation has been referred to by Berry (1932) with respect to black-tailed deer in Sonoma County, California.

Experimental Induction of Female Antlers

Whatever the cause might be, existence of antlered does in nature challenges the experimental zoologist to explore the possibilities of causing antlers to develop in females by surgical or hormonal means. Wislocki *et al.* (1947) succeeded in doing this in two female white-tailed deer by removing their ovaries and injecting testosterone. Both were spayed in January and given testosterone the following May or June, as a result of which pedicles developed but regressed by the end of the year. One of these does regrew pedicles and antlers the following summer after being injected with testosterone again in the spring. The other produced pedicles and short antlers without further treatment with testosterone. However, attempts to repeat this in the genus *Cervus* have not been as successful. In sika deer, injections of testosterone, or removal of the ovaries, have little effect when performed separately. However, in combination, both ovariectomy and testosterone treatment have induced conspicuous pedicles from the frontal bones of females (Fig. 151). On the one hand, these pedicles did not grow antlers from their tips during the course of several years' observations. On the other hand, they underwent annual fluctuations in length, becoming 2–3 cm high during the summer months, but regressing to about 1 cm each winter.

It was Jaczewski (1976, 1977, 1981, 1983), who discovered the secret of how to stimualte antler growth from such pedicles. First, he ovariectomized female red deer and injected them with testosterone to induce the development of pedicles. Second, he amputated the ends of these pedicles on one side, thus promoting the outgrowth of remarkably normal, but unilateral, antlers (Fig. 152). The pedicles on the unoperated sides remained unresponsive. This brilliant experiment yielded incontrovertible proof that female deer possess the latent capacity to grow antlers. It also called attention to the important role that injury and wound healing play in the initiation of antler regeneration. As in so many cases of spontaneous antler growth by female deer that have been reported, these experimentally induced antlers failed to shed their velvet. When the doe was given estrogen or testosterone, their

Fig. 151. Sika doe which had been spayed and treated with testosterone. Pedicles were grown which waxed and waned with the seasons. No antlers grew from them.

maturation was completed (Jaczewski, 1977, 1981, 1983). Casting and regeneration occurred when hormonal treatments were discontinued. Neither pregnancy nor lactation interferes with the induction of antler growth in female red deer (Jaczewski, 1981).

Bubenik et al. (1982) have reported the induction of an antler in a female white-tailed deer sterilized by previous treatment with antiestrogen compounds. Although this hormonal imbalance caused small pedicles to develop, they regressed following cessation of treatment. Several years later, a 4 × 10-mm piece of frontal periosteum was excised from the pedicle region through an incision in the scalp. This trauma, coupled with a similar operation a few months later, resulted in the outgrowth of a short velvet antler 3.5 cm high during the subsequent summer. This elongated to 12 cm the next year, but remained in velvet and froze in the winter. It was cast in January and replaced by a bifurcate regenerate during the following summer.

In another female, not treated with antiestrogens, trauma alone to the frontal periosteum also caused antler growth (Bubenik et al., 1982), indicating that prior hormonal conditioning may not be required to predispose does to antlerogenic induction by periosteal injury. This finding is consistent with the observations of Robbins and Koger (1981) that subcutaneous injections of $CaCl_2$ beneath the scalp of female wapiti causes extensive tissue necrosis that may result in antler production. Therefore, it would appear that female deer are endowed with a latent capacity for antler development susceptible to activation by traumatic stimulation.

Fig. 152. Female red deer with unilaterally induced antler. She had been ovariectomized and injected with testosterone to promote pedicle development. Amputation of the end of the left pedicle resulted in the production of a remarkably normal antler on that side. (Photo by Klaus-Jurgen Hofer; courtesy of Dr. Zbigniew Jaczewski.)

The foregoing experiments by Jaczewski and Bubenik involved the induction of antlers by female deer in the spring or summer, which is approximately when the males of these species ordinarily grow their antlers. However, the research of Robbins and Koger (1981) adds a new and interesting dimension to the phenomenon. They showed that female wapiti could grow antlers in the fall and winter in response to trauma inflicted between August and November. As in other examples of antlered does, these outgrowths remained in velvet. The explanation of how such antlers could be grown at the "wrong" time of year might lie in the fact that they were the first antlers produced by the deer. It is recognized that the initial sets of antlers grown by fawns or yearlings are not dependent on photoperiodic stimuli as are subsequent regenerates (Chapters 6 and 11). Traumatic induction of antlers in does may be similarly independent of the seasons, although later rounds of growth may not be.

There is a curious paradox with respect to antlered does. With very rare exceptions, such animals retain their antlers in velvet throughout the winter. However, it has been demonstrated experimentally that administration of estrogen may induce shedding of velvet (Goss, 1968; Jaczewski, 1977, 1981). Plotka et al. (1977a) reported relatively constant year round estrogen levels in nonpregnant white-tailed females, the very ones that Doutt and Donaldson (1960, 1961) found to be carrying bony antlers. Serum estrogen in pregnant does rises rapidly during the final 1–2 months of gestation, decreasing abruptly after parturition (Plotka et al., 1977b). Therefore, it is in pregnant antlered does that one might expect the velvet to be shed. In the autumn, the vicissitudes of estrogen production with each estrous cycle should also promote maturation of antlers in female deer. For reasons yet to be explained, none of these hormonal conditions induces velvet shedding from the antlers of female deer.

The Absence of Antlers in Males

The opposite of antlered does is the failure of males to produce the antlers they should. Although this is a specific characteristic of the musk deer and Chinese water deer, it occurs as an abnormality in white-tailed deer (Ryel, 1963; Seton, 1929), mule deer (Robinette and Gashwiler, 1955; Robinette and Jones, 1959), and in woodland caribou (Bergerud, 1976; Dugmore, 1913). In red deer, antlerless stags are known as hummels (Lincoln and Short, 1969). Although the absence of antlers in these deer might be a genetic character, the male offspring of hummels can and do in fact grow antlers (Lincoln and Fletcher, 1977; Lincoln et al., 1973). Though hummels are at an obvious disadvantage in competition with normal males in the herd, they are nevertheless completely fertile (Lincoln *et al.,* 1971). Hummels are reminiscent of polled cattle, that is, animals that have been bred to eliminate the production of horns. However, they sometimes possess short pedicles on their heads. Following the lead of Jaczewski (1976, 1983), Lincoln and Fletcher (1976) cut off the tip of one of these pedicles in the spring. This operation resulted in antler development on the amputated side. The antler grew to a length of 29 cm and developed 3 points. It shed the velvet and was cast the following spring, but subsequent regeneration could not be followed owing to the untimely death of the stag. Again, a healing wound is seen to be a key factor in triggering the onset of antler regeneration.

Further evidence along these lines comes from recent studies on how to prevent the production of antlers in male deer. Dehorning of cattle is usually achieved by the application of caustic agents to the horn bud region in calves. Comparable procedures in wapiti fawns have involved the subcutaneous injection of concentrated $CaCl_2$ (Robbins and Koger, 1981). This causes extensive necrosis of the antler-forming region of the head, and in some cases has the desired effect of preventing antler development altogether. However, as mentioned earlier, such procedures may exert a stimulatory influence in females, and can do likewise in males.

"Dehorning" of reindeer has been attempted by use of sodium hydroxide applied to the pedicle region 10 days after casting of the old antlers (Hadwen and Palmer, 1922). Although this treatment causes necrosis of tissues in the incipient antler, it does not necessarily prevent regeneration. In fact, it has been observed even to stimulate growth.

In white-tailed deer, Waldo *et al.* (1949) stopped the elongation of antlers with basal tourniquets applied in July, but noted that similar treatment earlier in the growing season was less effective, presumably because of changes in the susceptibility of the vascular pattern to the effects of constriction. Mautz (1977) tried to prevent antler growth by ligating early antlers with elastic bands. This caused them to become swollen, and later fall off, but renewed outgrowth sometimes occurred.

Attempts to inhibit antler development by chemical trauma, ischemic necrosis, or even surgical deletions (Bubenik and Pavlansky, 1965) all share the same disadvantage. They create lesions concomitant with the loss or destruction of tissue. Such lesions elicit healing reactions which are not only potential centers of renewed growth, but are also an indispensible component of the regeneration process. Even when full thickness skin is grafted over the end of the pedicle, the initiation of antler development is prevented only if special care is taken to reduce integumental wound healing to a minimum (Goss, 1972).

It is as difficult to prevent antler growth in males as it is to promote it in females. By analyzing the conditions under which antlers can be stimulated or inhibited where nature did not intend, basic principles of mammalian regeneration may eventually be revealed.

References

Alston, E. R. (1879). On female deer with antlers. *Proc. Zool. Soc. London* pp. 296–299.

Benirschke, K., Brownhill, L., Low, R., and Hoefnagel, D. (1963). The chromosomes of the white-tailed deer, *Odocoileus virginiana borealis*, Miller. *Mamm. Chromosome Newsl.* **10**, 82.

Bergerud, A. T. (1976). The annual antler cycle in Newfoundland caribou. *Can. Field Nat.* **90**, 449–463.

Berry, L. J. (1932). A "horned" blacktail doe with fawn. *J. Mammal.* **13**, 282–283.

Bubenik, A. B., and Pavlansky, R. (1965). Trophic responses to trauma in growing antlers. *J. Exp. Zool.* **159**, 289–302.

Bubenik, G. A., Bubenik, A. B., Stevens, E. D., and Binnington, A. G. (1982). The effect of neurogenic stimulation on the development and growth of bony tissues. *J. Exp. Zool.* **219**, 205–216.

Buechner, H. K. (1957). Three additional records of antlered female deer. *J. Mammal.* **38**, 277–278.

Buss, I. O. (1959). Another antlered female deer. *J. Mammal.* **40**, 252–253.

Buss, I. O., and Solf, J. D. (1959). Record of an antlered female elk. *J. Mammal.* **40**, 252.

Crispens, C. B., Jr., and Doutt, J. K. (1973). Sex chromatin in antlered female deer. *J. Wildl. Manage.* **37**, 422–423.

Diem, K. L. (1958). Fertile antlered mule deer doe. *J. Wildl. Manage.* **22**, 449.

Dixon, J. S. (1927). Horned does. *J. Mammal.* **8**, 289–291.

Donaldson, J. C., and Doutt, J. K. (1964). Antlers in female white tailed deer: A four year study. *Anat. Rec.* **148**, 366.

Doutt, J. K., and Donaldson, J. C. (1959a). Female Virginia deer with antlers. *Pa. Game News* **30**(11), 29–31.

Doutt, J. K., and Donaldson, J. C. (1959b). An antlered doe with possible masculinizing tumor. *J. Mammal. 40*, 230–236.

Doutt, J. K., and Donaldson, J. C. (1960). Female deer with antlers. *Pa. Game News* **31**(12), 52–53.

Doutt, J. K., and Donaldson, J. C. (1961). Antlered doe study. *Pa. Game News* **32**(11), 23–25.

Dugmore, A. A. R. (1913). "The Romance of the Newfoundland Caribou. An Intimate Account of the Life of the Reindeer of North America." Lippincott, Philadelphia.

Goss, R. J. (1968). Inhibition of growth and shedding of antlers by sex hormones. *Nature (London)* **220**, 83–85.

Goss, R. J. (1972). Wound healing and antler regeneration. *In* "Epidermal Wound Healing" (H. I. Maibach and D. T. Rovee, eds.), pp. 219–228. Year Book Med. Publ., Chicago.

Hadwen, S., and Palmer, L. J. (1922). Reindeer in Alaska. *U.S., Dep. Agric., Bull.* **1089.**

Hamilton, W. J., Harrison, R. J., and Young, B. A. (1960). Aspects of placentation in certain Cervidae. *J. Anat.* **94**, 1–33.

Haugen, A. O., and Mustard, E. W. (1960). Velvet-antlered pregnant white-tailed doe. *J. Mammal.* **41**, 521–523.

Jaczewski, Z. (1976). The induction of antler growth in female red deer. *Bull. Acad. Pol. Sci., Ser. Sci. Biol.* **24**, 61–65.

Jaczewski, Z. (1977). The artificial induction of antler cycles in female red deer. *Deer* **4**, 83–85.

Jaczewski, Z. (1981). Further observations on the induction of antler growth in red deer females. *Folia Biol. (Krakow)* **29**, 131–140.

Jaczewski, Z. (1983). The artificial induction of antler growth in deer. *In* "Antler Development in Cervidae" (R. D. Brown, ed.). Caesar Kleberg Wildl. Res. Inst., Kingsville, Texas (in press).

Kurnosov, K. M. (1962). Interfetal placental connections of the elk in embryonic parabiosis. *Dokl. Akad. Nauk SSSR* **142**, 253–256.

Lincoln, G. A., and Fletcher, T. J. (1976). Induction of antler growth in a congenitally polled Scottish red deer stag. *J. Exp. Zool.* **195**, 247–252.

Lincoln, G., and Fletcher, T. J. (1977). History of a hummel. Part 5. Offspring from father/daughter matings. *Deer* **4**, 86–87.

Lincoln, G. A., and Short, R. V. (1969). History of a hummel. *Deer* **1**, 372–373.

Lincoln, G. A., Guinness, F., and Short, R. V. (1971). The history of a hummel. Part 2. *Deer* **2**, 630–631.

Lincoln, G. A., Guinness, F. E., and Fletcher, T. J. (1973). History of a hummel. Part 3. Sons with antlers. *Deer* **3**, 26–31.

Mautz, W. W. (1977). Control of antler growth in captive deer. *J. Wildl. Manage.* **41**, 594–595.

Mierau, G. W. (1972). Studies on the biology of an antlered female mule deer. *J. Mammal.* **53**, 403–404.

Murie, O. J. (1935). "Alaska-Yukon Caribou," North Am. Fauna No. 54. U.S. Dept. Agric., Washington, D.C.

Plotka, E. D., Seal, U. S., Schmoller, G. C., Kerns, P. D., and Keenlyne, K. D. (1977a). Reproductive steroids in the white-tailed deer (Odocoileus virginianus borealis). I. Seasonal changes in the female. *Biol. Reprod.* **16**, 340–343.

Plotka, E. D., Seal, U. S., Verme, L. J., and Ozoga, J. J. (1977b). Reproductive steroids in the white-tailed deer (Odocoileus virginianus borealis). II. Progesterone and estrogen levels in peripheral plasma during pregnancy. *Biol. Reprod.* **17**, 78–83.

Robbins, C. T., and Koger, L. M. (1981). Prevention and stimulation of antler growth by injections of calcium chloride. *J. Wildl. Manage.* **45**, 733–737.

Robinette, W. L., and Gashwiler, J. S. (1955). Antlerless mule deer bucks. *J. Mammal.* **36**, 202–205.

Robinette, W. L., and Jones, D. A. (1959). Antler anomalies of mule deer. *J. Mammal.* **40**, 96–108.

Roux, J., and Stott, K. (1948). Antler-bearing by a female sika deer. *J. Mammal.* **29**, 71.

Ryel, L. A. (1963). The occurrence of certain anomalies in Michigan white-tailed deer. *J. Mammal.* **44**, 79–98.

Seton, E. T. (1929). "Lives of Game Animals." Doubleday, Doran & Co., Garden City, New York.

Waldo, C. M., Wislocki, G. B., and Fawcett, D. W. (1949). Observations on the blood supply of growing antlers. *Am. J. Anat.* **84**, 27–61.

Whitehead, G. K. (1971). Female deer with antlers. *Deer* **2**, 638.

Wislocki, G. B. (1954). Antlers in female deer, with a report of three cases in *Odocoileus*. *J. Mammal.* **35**, 486–495.

Wislocki, G. B. (1956). Further notes on antlers in female deer of the genus *Odocoileus*. *J. Mammal.* **37**, 231–235.

Wislocki, G. B., Aub, J. C., and Waldo, C. M. (1947). The effects of gonadectomy and the administration of testosterone propionate on the growth of antlers in male and female deer. *Endocrinology* **40**, 202–224.

Wurster, D. H., and Benirschke, K. (1967). Chromosome studies in some deer, the springbok, and the pronghorn, with notes on placentation in deer. *Cytologie* **32**, 273–285.

Medicinal Uses of Antlers

For reasons which can only be surmised, man has attached both mystical and medical significance to antlers and horns. The legend of the unicorn, for example, which seems to have arisen independently in more than one culture in ancient times (Hathaway, 1980), is as closely bound up with sexual innuendos as it is with pharmacologic fantasies. To capture this mythical beast, it was necessary to lure him with a virgin seated in the forest. The animal would supposedly lie beside her and place his head in her lap, whereupon he would fall easy prey to his captors (Fig. 123).

The value of unicorn horn was derived from its purported properties as a universal antidote to all the poisons feared by ancient monarchs vulnerable to assassination (Dove, 1936). Drinking vessels carved from such horns were highly prized. Even if the unicorn had really existed, it most surely would have become extinct to satisfy the demands of the marketplace. As it was, the narwhal tusk served as a convincing substitute. Fortunately its inaccessibility in arctic oceans not only saved the narwhal from extinction but maintained the high prices of its rare and fabulous tusks.

No less valuable today are the horns of the rhinoceros. Whether shipped to China as a medicine, powdered in India as an aphrodisiac, or carved into dagger handles by Yemeni tribesmen, the rhino's horn threatens its possessor with extinction at prices that doubled in only several years to $14,000 per pound in 1981. Worth more than their weight in gold, these horns have been responsible for the slaughter of thousands of African rhinos per year by poachers. One can only hope that when their numbers dwindle to levels insufficient to support the demands of the market, purveyors of rhinoceros horns will disappear as did the original unicorn merchants.

Historical Overview

The use of horns and antlers for medicinal purposes dates back to antiquity. The ancient Egyptians left records prescribing hartshorn, applied as an ointment, inhaled after burning or fumigation, or used as a powder with which to brush the teeth. However, it is in the Orient that the ancient traditions of folk medicine, based on the use of innumerable herbs prescribed in complex and arcane ways, have persisted to the present day.

These herbs are supplemented with selected animal products. Antelope horn from Mongolia and Siberia has been used to treat pulmonary and liver inflamation, convulsions, apoplexy, and rheumatism. But the antlers of deer are by far one of the most widely prescribed (and costliest) substances in Chinese and Korean pharmacology. Indeed, the differential between supply and demand has skyrocketed the price of velvet antler to such inflated levels that the production of antlers as a renewable resource has become a lucrative business for deer farmers. There are now hundreds of thousands of deer being farmed for their antlers.

In ancient times, deer were hunted as much for their antlers as their venison, a practice in China that may even have contributed to the extinction of Père David's deer in the wild state, possibly as late as historic times (Sowerby, 1954). In due course, the advantages of deer farming became obvious. Siberian stags were raised in Russia to supply the Chinese market, a source that has been supplemented by the domestic production of the sika, or spotted deer (Chen, 1973). The antlers of both of these species are highly valued for their medicinal properties in both China and Korea. Although other species are acceptable, they bring lower prices.

It is a curious coincidence that American Indians used the antlers of the moose to treat epilepsy, headache, vertigo, and snakebites (Merrill, 1920). Indeed, the "moose disease" that can cause paralysis, blindness, staggering, and eventually the death of the animal, was identified with human epilepsy by American Indians. They believed that the moose used its right hind hoof to pierce a vein in its left ear to effect a cure for this condition. Accordingly, such hooves, fashioned as rings or necklaces, eaten in powdered form, or inhaled as incense, were used to cure human epilepsy.

In southwestern United States and northern Mexico, the Cora Indians consume the velvet of deer antlers for contraceptive purposes. One wonders if these beliefs and practices could have been brought with the American Indians from their oriental origins. If so, then the medical traditions of the Orient may date back even to prehistoric times.

The Antler Business

Since the 1960s, deer farming has flourished in New Zealand (Yerex, 1979). This was a fortuitous turn of events because the population of red deer, originally imported from Great Britain to indulge the sporting preferences of British settlers, had outgrown even the hunting prowess of the New Zealanders. They had rapidly become such a national menace that they were being slaughtered in their hundreds from helicopters.

Enterprising farmers recognized the potentially profitable possibilities of turning this oversupply of deer to their advantage by filling the apparently insatiable demands of the oriental market, principally Korea, for velvet antlers. It was necessary only to fence off acreage, put out feed until the wild deer became accustomed to entering the enclosure, and then shut the gate at the right moment. Within a decade, deer farming in New Zealand grew to impressive proportions, relying mostly on red deer, but to a lesser extent on fallow deer and other species. Methods of efficient management have been developed with the assistance of the New Zealand Deer Farmers Association. As the price of red deer antlers doubled and tripled over several years in the late 1970s, more and more New Zealanders turned to farming deer to supply the oriental consumption which in Korea amounted to the importation of 20 tons of antlers in 1980 alone.

The advantages of raising deer domestically for this purpose are obvious. Unlike venison, antlers are a renewable resource because stags will grow new sets every year. Under favorable circumstances, a red deer stag can grow at least 10 pounds of velvet antlers each year. At prices in excess of $100 per pound, the farmer can realize a handsome profit on his investment. An equally profitable enterprise is run by the native Alaskans whose large reindeer herds supply quantities of antlers for the oriental market, thereby providing a rich source of income for the Alaskan reindeer herders (Luick, 1983). In more politically compatible times, Siberian deer farms raised maral stags and sika deer to satisfy the Chinese need for velvet antlers (Metyushev, 1963).

The dealer, upon inspecting the antlers produced by the deer farmer or herder, sorts them into several grades in accordance with his judgment of their quality for the market. Grade A antlers bring the highest prices. They are the ones that are perfectly shaped, have not been allowed to grow too long or become too calcified, and have been neatly and carefully removed from the deer. Their pore texture, as judged by the cut end, should be fine and even. The color should be pinkish, at least for the Korean market, as an indication of the blood content. Antlers may be graded down because of abnormal shapes, coarse or uneven porosity, excessive calcification as a result of belated amputation, pearlation of the beam, and damaged skin (defects of which are carefully sewn together).

Harvesting and Preparation

Methods of removing antlers vary. The ideal method is to sedate the deer for the operation. Alternatively, the animal can be immobilized, with a local anesthetic to deaden the pain. Unfortunately, animals are sometimes re-strained and the antlers cut off without sensory blockade. In any case, the antlers are best removed with a saw, although sometimes they are crudely amputated with bolt cutters. Unlike the Koreans, the Chinese favor antlers from which as much blood as possible has been drained. In order to mini-mize hemorrhaging from the deer, it is advisable to tie a tourniquet around the pedicle, or the base of the antler below the level of amputation. This prevents blood loss from the deer until the smooth muscles in the arteries constrict.

In general, the younger the antler the better. Calcification reduces the value of the antler, so the producer is faced with the dilemma of choosing between selling larger antlers at lower prices, or smaller ones at higher prices. In the sika deer, half-grown antlers are ideal, harvested when they still have only 2 points and the beginnings of a third. In red deer, they are allowed to grow considerably longer and to develop 3 to 5 points. They must be removed while the growing tips are still well rounded, prior to the development of the crown. Reindeer antlers should be about two-thirds grown. Mature bony antlers have little value in the medicine market.

In the case of the reindeer, amputated antlers may be preserved by freez-ing, sometimes packed in cotton to prevent subsequent damage in shipping. They must then be carefully dried before delivery to the retailer. Dried antlers may weigh only 30–40% of fresh ones. Drying is a carefully con-trolled process. The antlers must be heated enough to prevent them from rotting, but not so much as to cook them. On the tundra, reindeer antlers may be air-dried if the temperature is cold enough to prevent deterioration. Otherwise they are desiccated in a tarpaulin tent over a bed of coals (Yudin and Dobryakov, 1974). The temperature should be maintained between 50° and 90° C, sometimes for several days until the antlers become brittle enough to give just the right ring when knocked together. The desiccation process is more of an acquired skill based on the intuition of the producer than it is a precisely controlled technique.

In China, where deer farms have been maintained for at least 200 years to support the antler market, the freeze-dry methods of the Arctic cannot be applied in the summer months when antlers are ripe for harvesting. Here the unanesthetized deer is suspended in a specially designed cage to reduce struggling while the antlers are sawed off (Otway, 1981a,b). Blood may be removed by aspiration if desired. The antlers are then immersed repeatedly for brief periods of time in hot water before being dried for several days in a kiln. They are then air-dried to the appropriate consistency. Similar methods of intermittently cooking and heat-drying the antlers were adopted by the Russian producers (Metyushev, 1963).

The dried antlers require further preparation (Rennie, 1980). They are usually immersed in alcohol, to soften them slightly prior to being sliced with a hand guillotine into thin transverse sections anywhere from 25 μm to 2 mm thick. These dried sections are then ready for use in prescriptions.

a   b

Fig. 153. (a) A package of Manchurian velvet antler "chips" sold in Chinese drugstores as a blood tonic. (b) Contents of the envelope consisting of ersatz antlers sliced into very thin sections. The prescription is "4 chewable chips per dose."

They may be chewed as is, grated into a powder, stewed with meat for several hours in a soup, or even fried. Whatever the method of preparation, the concoction is taken internally, often in mixtures with various herbs, such as ginseng. But let the buyer beware. Slices of fake materials, cleverly fashioned to simulate real antlers, are sold at high prices in some Chinese drugstores to gullible consumers (Fig. 153).

The preparation of hartshorn in ancient Egypt was rather different from that in the Orient. There is an account of hartshorn, i.e., antlers of the stag, being mixed with "legs of a bird, the hair of an ass, and the dung of swallows and geese," a preparation used for such varied purposes as treating ulcers, uterine discharge, loose bowels, epilepsy, and even to get rid of snakes (Dawson, 1926). Spirits of hartshorn, or ammonia, is produced by distilling the shavings of antlers (Anthony, 1929).

Prescriptions

Despite the prevalent impression in the West that antlers are used in the Orient primarily as an aphrodisiac, the fact of the matter is that they are considered useful in treating a wide variety of medical problems. In addition to such specific conditions as epilepsy, ulcers, anemia, and rheumatism, antlers are prescribed for many lesser human ailments, including sleeplessness, tiredness, headache, lack of appetite, convulsions, and apoplexy (Yudin and Dobryakov, 1974). According to "The Great Dictionary of Chinese Medicine," deer horn, or lu rong (derived from the Siberian stag) is believed to be beneficial to the heart by slowing its beat, decreasing its size, and reducing blood pressure. It is also used to stimulate appetite, alleviate insomnia, improve kidney function, induce diuresis, and enhance wound healing in skin, ulcers, and bones. In addition to being used to treat fatigue, deafness and "cloudiness in the eye," lu rong has also been prescribed for such sexually related conditions as "shrinking penises," "slippery semen," "coldness in the uterus," "breakthrough bleeding," and "vaginal discharge." There would seem to be few conditions that do not respond to this ancient cure-all. Indeed, other parts of the deer are also used to treat various medical conditions. These include the bone, blood, testes, ligaments, tail, penis, and embryos. Whether such organs contain ingredients of proven medical usefulness, or exert their effects by the power of suggestion as broad spectrum placebos, has been the subject of only minimal scientific investigation. Whatever the truth of the matter, the fact remains that for centuries, if not millennia, these treatments have been the cornerstone of traditional oriental medicine (Chen, 1973).

Scientific Investigations

The active ingredient of deer antlers, although neither isolated nor purified, has been given the name of pantin or pantocrin (Chen, 1973). The extract from reindeer antlers is called rantarin (Yudin and Dobryakov, 1974). Although a number of investigations on experimental animals have been carried out in efforts to confirm the reported efficacy of pantin or rantarin, these studies have not always been satisfactorily controlled.

The supposed aphrodisiac qualities of deer antler have suggested to some that testosterone could be responsible for the effects. On the one hand, antler extracts have been found not to stimulate the growth of seminal vesicles in castrated rats, an effect which would have been expected from testosterone (Yudin and Dobryakov, 1974). On the other hand, when velvet was fed to growing chickens, there was an increase in spermatogenesis and testosterone secretion in the testes (Bae, 1975). In mice, rantarin has been observed to promote precocious sexual development, suggestive of a gonadotrophic effect (Yudin and Dobryakov, 1974).

Although consumption of antler extract by human beings has proven that it has no toxic effects, it has been shown to promote oxygen metabolism in the brain, liver, and kidneys of mice. It may also exert an influence on blood cell production. Hemoglobin and red blood cell counts were increased in rabbits treated with velvet antlers, and alcoholic extracts increased erythropoietin in the plasma and iron uptake in the hemoglobin of rabbit red cells (Song, 1970). It also enhanced hemoglobin production and increased the hematocrit of chickens (Bae, 1976). It is said to lower the blood pressure (Yudin and Dobryakov, 1974). Intravenous injections of alcoholic extracts of Korean deer antlers was found to increase cardiac stroke volume in dogs, but exerted no significant effects on heart rate, cardiac output, or various blood pressure parameters (Clifford et al., 1979).

Some studies have focussed on the possibility that antler extracts might affect physiological reactions to stress. Accordingly, some of the effects of stress on experimental animals have been reduced by antler extract, including adrenal hypertrophy, involution of the thymus, gastric hemorrhage, decreased liver and kidney weight, and the fall in enterochromaffin cells (Han, 1970; Yudin and Dobryakov, 1974). In human beings, antler extracts have been reported to improve strength and intellectual activity. In pigs (Bae, 1976) and chickens (Yudin and Dobryakov, 1974), accelerated weight gains have been noted in response to feeding velvet antlers.

A note of caution is in order. Some experiments designed to explore the possible physiological effects of antler have not been conducted with adequate controls. Further, there is a tendency in scientific circles not to publish negative evidence, and it can be assumed that an undetermined number of studies may have gone unpublished because they did not yield positive results. This is not to say that velvet antlers are without physiological influences. According to the standards of western science, they must be presumed ineffectual until proven and confirmed by carefully controlled experiments. In the meantime, the ancient practice of using deer antlers in traditional oriental medicine will continue to provide a market for the many antlers grown by deer every year. It is convenient if the beliefs of mankind are supported by scientific evidence. Lacking this, they seem to go on being believed anyway.

Ethical Pros and Cons

What about the moral implications of the practice of using deer antlers in medicine? Whether the substance works or not, the fact remains that it is a firmly entrenched part of oriental medicine. However, it is difficult to assess the extent to which scientifically proven treatments may be neglected in favor of more traditional practices.

From the point of view of the deer, humane considerations would dictate that if growing antlers are to be removed, the operation should be done as painlessly as possible. Antlers are supplied with numerous sensory nerve fibers, and there is every reason to believe that amputation of an antler must be at least as painful to the deer as would be the loss of any other of its appendages. Accordingly, appropriate veterinary supervision to ensure that the sensations in the base of the antler are adequately blocked is of obvious importance.

Fortunately, there seems to be no danger that deer species will become extinct because of the demand for their antlers. Although animals are undoubtedly still hunted for this purpose, the lion's share of velvet antlers derives from deer farming operations that do not necessitate killing the animals in order to obtain their antlers. However, it cannot be denied that this practice probably encourages the continued demand for other products, such as rhinoceros horns. Although these horns are said to be capable of continued regrowth after their removal, for obvious reasons it is not feasible to obtain them without killing the rhino, a practice that promises eventually to abolish the trade by extinguishing the source.

There is one important advantage of the antler trade, aside from its financial rewards. This relates to the enhanced popularity of deer farming which in turn generates increased interest in deer and their antlers. Therefore, a by-product of the business should be an increased recognition of deer antlers as subjects of serious scientific investigation. With cooperation between the traditional medical establishment in the Orient, deer farmers around the world, and the research scientist, answers to some of the unsolved problems of antler biology may eventually be forthcoming.

References

Anthony, H. E. (1929). Horns and antlers. Their occurrence, development and function in the Mammalia. Part II. *Bull. N. Y. Zool. Soc.* **32,** 3–24.

Bae, D.-S. (1975). Study on the effects of velvet on growth of animals. I. Effects of velvet of different levels on weight gain, feed efficiency and development of organs of chicken. *Korean J. Anim. Sci.* **17,** 571–576.

Bae, D.-S. (1976). Studies on the effects of velvet on growth of animals. II. Effects of velvet on the growth of internal organs and blood picture of chicken. *Korean J. Anim. Sci.* **18,** 342–348.

Chen, J. Y. P. (1973). Chinese health foods and herb tonics. *Am. J. Chin. Med.* **1,** 225–247.

Clifford, D. H., Lee, M. O., Kim, C: Y., and Lee, D. C. (1979). Can an extract of deer antlers alter cardiovascular dynamics? *Am. J. Chin. Med.* **7,** 345–350.

Dawson, W. R. (1926). Studies in ancient materia medica. *Am. Drug.* **4,** 17–18.

Dove, W. F. (1936). Artificial production of the fabulous unicorn. A modern interpretation of an ancient myth. *Sci. Mon.* **42,** 431–436.

Han, S. H. (1970). Influence of antler (deer horn) on the enterochromaffin cells in the gastrointestinal mucosa of rats exposed to starvation, heat, cold and electric shock. *J. Cathol. Med. Coll., Seoul* **19,** 157–166.

Hathaway, N. (1980). "The Unicorn." Viking Press, New York.

Luick, J. R. (1983). The velvet antler industry. *In* "Antler Development in Cervidae" (R. D. Brown, ed.). Caesar Kleberg Wildl. Res. Inst., Kingsville, Texas (in press).

Merrill, S. (1920). "The Moose Book." Dutton, New York.

Metyushev, P. V. (1963). History, economic significance, and husbandry. The deer used in the antler industry. *In* "Deer Husbandry," by I. V. Druri and P. V. Metyushev. Publ. House Agric. Lit., Moscow; Engl. transl. in *Reindeer Herders Newsl.* **7**(1), 1–30 (1982).

Otway, W. (1981a). China's deer industry. *N. Z. Farmer* **102**(22), 17–19.

Otway, W. (1981b). China's deer industry. The Chinese view of velvet quality and use. *N. Z. Farmer* **102**(23), 18–19.

Rennie, N. (1980). Niel Rennie on velvet. *N. Z. Farmer* July 10 (pp. 8–9), July 24 (pp. 20–23), Aug. 14 (pp. 10–12), Aug. 28 (pp. 27–31), Sept. 11 (pp. 26–28), Sept. 25 (pp. 20–23), Oct. 9 (pp. 28–29).

Song, S. K. (1970). Influence of deer horn on erythropoietin activity and radioactive iron uptake in rabbits. *J. Cathol. Med. Coll., Seoul* **18,** 51–60.

Sowerby, A. de C. (1954). The range of Father David's deer. *Anim. Kingdom* **57,** 83–85.

Yerex, D. (1979). "Deer Farming in New Zealand." Agric. Prom. Assoc., Wellington.

Yudin, A. M., and Dobryakov, Y. I. (1974). "Reindeer Antlers. A Guide for the Preparation and Storage of Uncalcified Male Antlers as a Medicinal Raw Material." Acad. Sci. USSR., Far East Sci. Cent., Vladivostock.

Index

A

Abnormalities
  of antlers, 4, 27, 129–131, 138, 142,
    143, 145–148, 167, 193–212, 236,
    237, 242, 243, 259, 262, 268–281,
    285–287, 289, 300
  of horns, 212–215
Acidophil, 256, 257
ACTH, 256
Adrenal gland, 257, 258, 281
  hypertrophy, 174, 258
  tumor, 288
Adrenalin, 166
Africa, 8, 74–76, 92–94
Age, 107, 108, 197, 198, 237, 238
  determination, 60, 86, 212
Alkaline phosphatase, 158, 160, 161
Allometry, 31, 89, 90, 97, 114, 118
β-Aminopropionitrile, 261
Amputation, 107, 108, 140, 142, 161, 165,
    167, 174–177, 179–182, 189, 190,
    201, 204–206, 209, 221, 255, 290,
    294, 300, 304
Annual rings, 60, 64, 65
Antelope, 54–56, 58, 62–65, 93, 117, 298
Antiandrogen, 210, 269
Antiestrogen, 291
Antilocapridae, 77
Antler
  bud, 3, 97, 124, 125, 146–148, 150,
    151, 153–158, 205, 206, 253

cycle, 15, 19, 28, 36, 38, 44, 231–246,
    252, 260
  in female, 4, 33, 34, 116, 209, 210, 255,
    284–293
  first, 3, 9, 12, 28, 30, 32, 33, 116,
    122–131, 249, 250, 253, 287, 293
  function, 103, 104, 106–119, 181, 218,
    222, 224
  fused, 27, 194, 195
  histogenesis, 3, 127–131, 141–147
  infant, 32, 33, 116, 125, 126
  lack of, 4, 26, 35, 118, 284, 293–295
  locked, 106
  onset of growth, 138, 139, 141, 142,
    148, 231, 234, 253, 260, 271
  orientation, 211
  peruke, 198, 276–278
  retention of, 14, 39, 235–238, 242, 252
  series, 224, 225, 248, 249
  size, 98, 107, 108, 114, 134, 208–211,
    236–238, 260, 262, 263
  strength, 107, 114
  transplantation, 146, 147
  water content, 107, 161
Antlerogenic transformation, 124, 125
Antleroma, 4, 212, 272, 274–281
"Antlerotrophic" hormone, 256, 257, 281
Aphrodisiac, 4, 297, 302, 303
Apical epidermal cap, 151
Apophysis, 67, 124
Arctic, 33, 238, 296, 300, 301
Aristotle, 127, 137, 268

Arrector pili muscle, 149, 166
Art, 8, 10, 11, 92–94, 214, 215
Arteriovenous anastomosis, 164
Artery, 163–166, 204, 261, 268
  carotid, 163
  constriction, 165, 166, 177, 204, 300
  tail, 165
  temporal, 163
  umbilical, 165
Asia, 8, 11–27, 44, 85–87, 93, 115, 241, 254
Asymmetry, 33, 34, 69, 70, 194, 200, 218–228, 285, 289, 290, 292
Asynchronous antler cycles, 14, 18–21, 24, 39, 41, 230
Autotomy, 141, 165, 175–177, 206
Axis deer, 7, 20, 21, 44, 115, 242, 272

B

Babirusa, 52, 69
Barasingha, 14, 15, 44, 46
Barbary stag, 92
Barking deer, 24
Barren females, 116, 255, 285
Basal snag, 37, 96
Base of antler, 140, 271
Bat, 187, 188, 190, 191, 220
  wing membrane, 187, 188, 190, 191
Bedford, Duke of, 21–23, 262
Behavior, 2, 7, 10, 35, 101–108, 116, 117, 231, 240, 251, 259
Bell, 31
Birth, 7, 9, 10, 12, 14, 15, 17, 18, 21, 22, 24, 26, 28, 30, 31, 33, 35, 39–41, 44, 113, 230, 233, 238–241, 254–256, 261, 287, 293
Black buck, 55, 58
Black-tailed deer, 35–39, 162, 269, 290
Blastema, 2, 3, 148, 151, 177–181, 183, 184, 186, 191, 206
Blastomeryx, 95
Blind deer, 259, 260
Blood flow, 109, 110, 115, 124, 142, 143, 148, 153, 155, 156, 159, 160, 162, 163, 201, 203, 207, 226, 269, 288, 289, 294
Blood vessels, 127, 133, 141, 142, 143, 148, 153–159, 163–165, 179, 182, 280, 294

Bone, 66, 82, 83, 107, 123, 124, 127, 131, 142, 147, 154, 155, 157, 160–162, 261, 279
  core, 52, 57, 64, 67, 81, 124, 213
  density, 262
  frontal, 67, 76, 114, 122, 123, 127, 129, 142, 147, 206, 250, 253, 290
  metacarpus, 44, 45, 95, 130
  parietal, 67, 68, 130
  turnover, 162
  water content, 107, 161
Bovidae, 6, 54–62, 65–67, 75, 124
Brocket, 39, 40, 44, 46, 96, 241, 254
Broken antlers, 107, 108, 114
Brooming, of horns, 65
Brow-antlered deer, 16, 17
Brow tine, 10, 16, 22, 34, 74, 89, 90, 106
  angle of, 222, 227
  asymmetry of, 218, 219, 222–228
  lack of, 22, 23, 37, 96, 145, 203, 206, 223
Bugling, 8, 10, 29, 104
Burr, 74, 76, 78, 82, 85, 86, 138, 272, 280, 287, 289
Bush-antlered deer, 87, 88, 97

C

CaCl$_2$, 66, 291, 294
Cactus buck, 274
Cadmium, 261, 269
Calcification, 154, 155, 158, 300
Calcium, 90, 91, 133, 155, 161, 262
Cancer, 4, 133, 173, 212, 215, 278–281, 288
Caribou, 1, 33–35, 44, 95, 103, 106, 107, 109, 116–118, 125, 135, 138, 163, 166, 218–220, 222–228, 230, 249, 251, 254, 255, 270, 271, 277, 278, 284, 285, 287, 293
Cartilage, 67, 68, 124, 154, 155, 158, 161, 164, 178, 179, 181, 183–187, 254, 261, 279
  calcified, 155
Cartilaginous sheet, 183–186
Cartilaginous zone, 153, 155
Casting of antlers, 9, 10, 12, 14, 17, 23, 28, 30, 32–36, 39, 40, 42, 44, 85, 107, 111–116, 137–141, 144, 147, 167, 237–242, 251–254, 278, 285, 287, 291

delayed, 14, 21, 39, 235–238, 242, 252, 259, 262, 278, 285
  failure of, 138, 199, 207, 221, 231, 233, 234, 242, 253, 255, 258, 287
  premature, 243, 262, 271
Castration, 4, 33, 64, 65, 107, 115, 127, 198, 211, 212, 242, 243, 255, 257, 261, 268–281, 287, 303
  unilateral, 203, 270
Cattle, 55, 66, 215, 288, 294
Cave paintings, 10, 11, 92, 93
Cervalces, 87, 88
Chameleon, 52, 53, 177
Chevrotain, 25
China, 11, 21–23, 24, 26, 44, 86, 297, 298, 301
Chinese water deer, 26, 27, 29, 44, 46, 95, 293
Chondroblast, 153
Chondroclast, 124, 158
Chondrocyte, 155, 157, 158
Chondrogenesis, 123, 154–158, 183, 279
Chromosome, 23, 43, 44, 46, 95, 96, 288, 289
Circadian rhythm, 243, 246
Circannual rhythm, 4, 199, 235, 238, 243–246, 260
Classification of deer, 42–47, 95
Climacoceras
  africanus, 74, 75, 94
  gentryi, 74, 76
Climate, 86, 87, 91, 92, 109, 114, 115, 118, 230, 231, 240, 260, 272
Collagen, 148, 152, 153, 155, 158, 161, 173, 183, 191, 261, 279
Collagenase, 191
Collateral circulation, 163
Color, 9, 11, 12, 14, 20, 22, 26, 29, 32, 36, 115, 117, 231, 243, 259
Compensatory regulation, 221
Constant illumination, 199
Contraception, 299
Contralateral effect, 201–203, 208, 270
Corkscrew antlers, 197, 198
Cornification, 52, 57, 63, 64, 72, 81, 82, 212, 213
Corpus luteum, 254, 256
Cortisone, 191, 256, 258
Cosoryx, 77, 78
Cross-breeding, 8, 10, 33, 36, 95

Crown, 10, 300
Cryptorchidism, 203, 269, 288
Cyproterone acetate, 269

D

Dama gazelle, 55, 56
David, Père Armand, 21
Day length, 4, 126, 127, 199, 202, 230–238, 240, 242–246
Dedifferentiation, 178, 191
Deer farming, 4, 10, 204, 298, 304
Dehorning, 66, 213, 294
Dehydrogenase, 160
Delayed antler growth, 14, 32, 34, 36, 39, 138, 142, 199, 235–238, 242, 252, 253, 259, 262, 271
Delayed casting, 14, 21, 39, 235–238, 242, 252, 259, 262, 278, 285
Delayed implantation, 22, 28, 29, 254, 256
Deletion
  of bone, 66, 127, 128, 142, 206
  of cartilage, 184, 185
  of periosteum, 66, 130
  of skin, 66, 127, 128, 131, 142, 144
Denervation, 167, 208
Dermis, 142, 147, 148, 152, 153, 155, 160, 173, 183, 279, 280
Dewclaw, 45, 95
Dicroceros, 85, 97
Differentiation, 125, 141, 147, 149, 153, 159, 173, 180, 184, 187, 250, 279
Display, 87, 89, 105–110, 116, 117, 222
Diversity of deer, 6–51
Domestication, 32, 33, 299
Dominance, 104, 106–108, 117
Double
  antlers, 196, 197, 205, 206
  horns, 213
Dybowski's deer, 12

E

Ear, 143–146
Eating of antlers, 103, 138, 139
Economic value of antlers, 4, 10, 33, 299, 300
Ectopic
  antlers, 129, 130, 145, 148, 195, 206, 207
  horns, 66, 213–215

Edema, 142, 143, 163, 207, 294
Egypt, 92–94, 298, 302
Elaeophora schneideri, 259
Elastration, 269
Eld's deer, 16, 17, 44
Electric stimulation, 208
Elephant, 52, 69, 86, 155
Elk
  American, *see also* Wapiti, 29–31, 44,
    102
  European, *see also* moose, 31, 44
Ellesmere Island, 230
Elongation, 64, 70, 125, 153, 154, 208,
  231, 238, 258, 291
  rate, 3, 58–60, 69, 90, 133–135, 137,
    167, 212, 236, 250, 251
Eocene epoch, 72
Epidermal downgrowths, 150, 151, 183,
  184, 186
Epidermis, 52, 57, 58, 68, 72, 127, 142,
  144, 145, 147, 148, 150–153, 166,
  173, 178, 179, 181–184, 186, 189,
  191, 206, 210
Epididymis, 270
Epilepsy, 298, 302
Epineurium, 166, 167
Epiphysis, 67, 68, 124, 155
Equator, 19, 126, 238, 240, 242–246
Equinox, 232, 233, 237, 238, 240, 244,
  246
Estrogen, 127, 250, 254, 255, 278, 290,
  293
Estrus, 7, 9, 10, 24, 31, 33, 60, 105, 118,
  240, 254, 255, 293
  postpartum, 24, 38, 39, 241, 254
Eucladoceros (Euctenoceros), 87, 88
Evolution, 2, 8, 29, 35, 42–47, 52, 54, 62,
  63, 72–100, 108–118, 179, 181, 187,
  190, 194, 230
  of antler casting, 113–116
  of antler function, 108–110
  of velvet shedding, 111, 112
Extinction, 17, 18, 21, 40, 41, 73, 90, 91,
  92, 297, 298
Extracts of antler, 4, 303

F

Fallow deer, 7, 8–10, 44, 47, 92, 93, 103,
  106, 109, 110, 112, 123, 129, 130,

139, 146, 150, 161, 194, 201, 251,
  258, 270, 274, 279, 280, 299
Fawn, 7, 30, 36, 113, 115, 118, 122–131,
  161, 242, 249, 250, 253, 255, 261,
  268, 270, 288, 294
Fertility, 21, 24, 36, 39, 60, 115, 236, 241,
  243, 249, 250, 252, 254, 269,
  287–290, 294
Fetus, 67, 68, 116, 122, 262, 288
Fibroblast, 153, 155, 279
Fighting, 10, 20, 23, 26, 28, 29, 31, 54,
  60, 70, 84, 89, 101–108, 110–112,
  114, 118, 166, 222
First antler, 3, 9, 12, 28, 30, 32, 33, 116,
  122–131, 249, 250, 253, 287, 293
Flehmen, 105
Follicle stimulating hormone, 257
Fossil, 35, 42, 46, 63, 72, 75–97, 109,
  114, 115
Fracture, 173, 201, 203, 204
  nonunion, 203
Freemartin, 288, 289
Freezing, 76, 84, 107, 114, 115, 204, 212,
  272–274, 276, 278, 287, 289, 291,
  300
Function of antlers, 103, 104, 106–119,
  181, 218, 222, 224
Fur-covered horns, 73–76
Fused antlers, 27, 194, 195

G

Gall bladder, 25
Gemsbok, 54, 55
Genetics
  of abnormalities, 193–196, 212
  of antler morphology, 97, 248, 249
  of antler size, 97, 211, 212, 260–263
  of asymmetry, 218, 220, 221, 225
  of coat color, 9
  of female antlers, 284, 287–290
  of hummels, 293
Geographic distribution, 8, 42, 44, 46,
  84–96, 230, 238–240
Gestation, 7, 9, 10, 14, 15, 21, 22, 24–29,
  31, 33, 36, 39, 42, 250, 254, 256,
  259, 293
Giant deer, 87–93, 109, 118
Ginseng, 302
Giraffe, 6, 52, 67–69, 73–76, 92, 124

Gland, exocrine
facial (preorbital), 7, 22, 103
interdigital, 7, 110
mammary, 25
metatarsal, 7, 110
musk, 7, 25
sebaceous, 110, 124, 148, 149, 166
tarsal, 7, 110
Glycogen, 158
Gnu, 55
Goat, 67, 215
Gonadotrophic hormone, 253, 256, 257, 258, 259
Grain of velvet, 150
Grooming, 203
Growth
compensatory, 173, 174
curve, 133, 134
hormone, 256, 257
ring, 54–62, 135, 137, 212
zone, 148, 153–161, 164, 182, 184

H

Hair, 7, 124, 143, 144, 145, 148–150, 173, 203, 259
follicle, 3, 124, 148–150, 166, 184, 254
Harem, 22, 103, 104
Hartshorn, 298, 302
Haversian system, 140, 162
Hearing, 7
Hemorrhage, 165, 204, 300, 303
Herd, 33, 116–118, 222, 285
Hermaphrodite, 287
Hibernation, 231
Histamine, 166
Histogenesis
of antler, 3, 127–131, 141–147
of antleroma, 279, 280
of horn, 66–69
Hog deer, 19, 20, 44
Holocene epoch, 92
Hoof flailing, 102
Horn, 2, 52–71, 72–76, 117, 131, 135, 212, 285, 294, 297, 298, 304
bud, 57, 58, 66, 215, 294
fusion, 215
growth, 55, 57–60, 63, 64, 66–69, 81, 124, 135, 212
in man, 214, 215

rings, 54–62, 212
sheath, 52, 57, 58, 62, 63, 67, 72, 80, 82, 84, 212
Huemul, 41, 43, 44, 96, 231
Hummel, 4, 194, 261, 263, 293
Hunter, John, 163
Hunting, 1, 35, 194, 260, 262, 285–287, 289, 298, 299, 304
Hypertragulidae, 73
Hypophysectomy, 256
Hypothalamus, 280

I

Ibex, 60, 61
Ice age, 92, 95, 118
Impala, 64
India, 13, 14, 19, 21, 23, 44, 65, 74, 86, 87, 93, 297
Indian, American, 298, 299
Indonesia, 19, 23, 44
Induction, 123–125, 129–131, 141, 145, 191, 204, 206, 210, 280
Infant antlers, 32, 33, 116, 125, 126
Inhibition of antler growth, 281, 295
by $CaCl_2$, 294
by hypophysectomy, 256
by light cycle, 242
by NaOH, 294
by skin graft, 148
by testosterone, 253
by tourniquet, 294
Injury, 106, 107, 109, 111, 114, 160, 167, 183, 191, 193, 195, 200–210, 212, 249, 255, 280, 290, 291–293
contralateral, 200–203
Invertebrates, 174–176, 181, 219, 221
Irish elk, see also Megaceros, 2, 87, 89–93, 97, 106, 109, 118
Irregular antler cycles, 243
Ischemia, 140, 144, 146, 268, 269, 279, 294

K

Keloid, 280
Keratin, 52, 57, 58, 64, 65, 183, 212
Kerguelen Island, 232
Key deer, 35, 36
Kinesthetic sense, 166, 203

Korea, 299, 300
Kudu, 55

L

Lactation, 33, 287, 291
Lagomeryx, 76
Lathyrism, 261
Latitude, 4, 35, 36, 39, 44, 47, 84, 86, 93,
    96, 110, 118, 125, 225, 226, 230,
    234, 238–246
Lengths of antlers, 10, 12, 14, 16, 27, 29,
    31, 34, 37, 39–42, 87, 90, 262, 285
Leydig cells, 250, 269
Ligation, 142, 161, 163, 164, 268, 294,
    300
Locked antlers, 106
Lung worm, 197
Luteinizing hormone, 257

M

Malnutrition, 115, 190, 260–263
Markhor, 55, 59
Marsh deer, 41, 42, 44
Medicinal properties, 4, 10, 21, 297–305
Megaceros, 87, 89–93, 106, 118
Melatonin, 258, 259
Meryceros, 78–80
Merycodontidae, 2, 77–84
Merycodus, 78, 81
Mesoderm, 145–147, 150, 151, 178, 183,
    191
Mesopotamian deer, 9, 93
Metabolism, 91, 109, 133, 155, 161, 190,
    258, 261, 303
Metachromasia, 158
Metaplasia, 158, 179
Metastasis, 215, 278, 281
Migration
    animal, 29, 30, 33, 41, 47, 73, 96, 118,
        222, 231
    cell, 144, 146, 147, 173, 178, 181–183,
        189, 191
Mi-lu, 21–23
Mineral, 90, 91, 133, 138, 161, 162, 231,
    251, 252, 260, 261, 269
Miocene epoch, 62, 72–77, 84–87, 94,
    95, 97, 115
Molt, 7, 11, 14, 26, 29–33, 36, 38, 65,
    150, 221, 231, 243, 256, 259, 260

Moose, 1, 7, 31–33, 44, 46, 47, 87, 89,
    95, 103, 106, 109, 118, 125, 134,
    137, 139, 165, 166, 194, 206, 249,
    251, 254, 269, 271, 285, 289, 298
    disease, 298
Morphogenesis, 151, 167, 168, 178, 180,
    205, 206, 281
Moschidae, 46
Mouse, 130, 131, 194, 190
    nude, 130, 131
Mucopolysaccharide, 158
Mule deer, 35–39, 44, 103, 104, 106, 125,
    195, 197, 208, 238, 239, 260, 269,
    270, 272, 285, 287–289, 293
Muntjac, 23, 24, 44, 46, 85, 103, 241,
    249, 254
    Indian, 23, 46
    Reeves', 23, 24, 46
Musk deer, 7, 25, 26, 44, 46, 293
Mutation, 194, 211
Myelination, 166

N

NaOH, 294
Narwhal, 54, 69, 70, 219, 297
Necrosis, 9, 66, 111, 114, 115, 140, 143,
    146, 164, 268, 269, 272, 273, 276,
    278, 279, 281, 291, 294
Neoteny, 117
Nerve, 127, 133, 142, 148, 164, 166–168,
    176, 179, 180, 182, 191, 200, 203,
    208–210, 226, 280, 304
    supraoptic, 166
    trigeminal, 166
Neuroma, 167
Neurotrophic influence, 167, 176, 179,
    180, 191, 208
New Zealand, 10, 232, 299
Nilgai, 65
Nondeciduous antler, 77–84
Nutrition, 116, 125, 190, 198, 231, 248,
    260–263

O

Odocoileus, 35–39, 47, 95, 96, 106, 211,
    238–240, 249, 251, 254, 271
Okapi, 53, 73–75
Olfactory projection, 110
Oligocene epoch, 72, 73, 75

Onset of antler growth, 138, 139, 141, 142, 148, 231, 234, 253, 260, 271
Os cornu, 67, 75, 124, 131
Ossicone, 67–69, 75, 124
Ossification, 67, 68, 111, 112, 130, 154, 158–162, 164, 200, 231, 251, 253, 272, 278, 287
  endochondral, 158
Osteoblast, 158–160
Osteoclast, 115, 140, 271
Osteocyte, 157, 158
Osteoporosis, 162
Ovary, 220, 231, 254, 255, 287, 290, 291

P

^{32}P, 160
Pair bond, 117, 118
Palaeomerycidae, 75, 76
Palaeomeryx, 76
Palmate antlers, 1, 9, 31, 32, 34, 85–93, 106, 109, 134, 137, 194, 196, 218, 219, 222–228
Pampas deer, 40, 41, 44, 96
Pantin (Pantocrin), 303
Paracosoryx, 77, 83
Parallel walk, 105, 106
Parathyroid hormone, 140
Pearlation, 37, 160, 277, 300
Pedicle, 9, 23, 25, 27, 36, 46, 77, 82, 85, 97, 114, 122–125, 127, 129, 130, 138, 140–148, 160, 161, 163, 194, 199, 200, 201, 206, 207, 209, 211, 237, 242, 250, 253, 255, 256, 260, 271, 274, 288, 290–292, 294
Père David's deer, 21–23, 28, 44, 47, 103, 106, 110, 113, 211, 212, 262, 298
Perichondrium, 153, 155, 158
Periosteum, 3, 66–68, 75, 123, 127, 129, 130, 131, 141, 145, 147, 152, 153, 155, 160, 205, 210, 250, 279, 291
Peruke, 198, 276–278
Pheromone, 7, 105, 110, 124
Phosphorus, 133, 160, 161, 262
Photoperiod, 4, 126, 127, 138, 198, 199, 202, 230–238, 240, 242–246, 248, 257–259, 280, 293
  lengthened, 235, 237, 238
  shortened, 231–237
Pinealectomy, 258–260
Pineal gland, 258–260, 280

Pituitary, 256–259, 280
Placental fusion, 288, 289
Pleistocene epoch, 30, 87, 88, 91, 92, 95, 97, 118
Plesiometacarpalia, 44–46, 95
Pliocene epoch, 35, 74, 85, 87, 92, 95, 96, 118
Points, number, 10–12, 14–16, 18, 19, 21, 37, 40, 41, 85–87, 211, 260, 262, 263, 270, 271
Polled, 213, 294
Potash, caustic, 66
Predation, 106, 115, 116, 118, 175–177, 190
Pregnancy, 116, 255, 256, 262, 285, 287, 288, 291, 293
Preosseous mesenchyme, 157
Prescription, 302
Progesterone, 254–256
Prolactin, 257, 259
Proliferation, 57, 147, 150, 153, 155, 156, 163, 173, 174, 178, 182
Pronghorn antelope, 6, 62–65, 81, 115, 212
Prosynthetoceras, 73
Protein, 262, 263
Protoceratidae, 73
Pseudoburr, 77–84
Puberty, 9, 27, 29, 31, 36, 65, 101, 249, 250, 260, 270
Pudu, 1, 42–44, 96

R

Radioactivity, 162
Ramoceros, 78
Rantarin, 303
Red deer, 1, 8, 10, 11, 29, 30, 44, 47, 92, 95, 97, 102, 104–106, 110, 118, 125, 127, 134, 135, 146, 162, 195–197, 200, 203, 204, 209, 231, 248, 249, 251, 253–255, 257, 260, 262, 270, 285, 288, 290–293, 299, 300
Regeneration, 2, 3, 111, 114, 123, 133–168, 172–191, 208, 221, 263, 295
  of amputated antler, 204–206, 272–274, 287, 291
  of barbel, 176
  of bat wing, 187, 188, 190, 191
  of ear, 151, 181–187, 190, 191

Regeneration (*continued*)
  of eardrum, 187
  epimorphic, 148, 177–190
  of fin, 176
  of fingertip, 189–191
  of limb, 175, 176, 178, 179, 190, 191
  in mammals, 180–191
  of mouse digit, 190
  physiological, 172, 174
  of tail, 176, 177, 179, 181
  territory, 142, 178, 206
Regression (of pedicles), 290, 291
Reindeer, 1, 33–35, 44, 46, 47, 92, 95,
    106–109, 115–118, 125, 138, 148,
    160, 163, 194, 204, 218–220,
    222–228, 230, 249, 251, 254, 255,
    257, 258, 262, 270–272, 277, 278,
    284, 285, 287, 294, 299, 300, 303
Reproductive cycle, 7, 14, 17, 29, 36, 38,
    39, 41, 44, 115, 230, 234, 238, 239,
    248, 258–260
Reserve mesenchyme, 153
Reversed
  asymmetry, 221
  polarity, 180, 206
  seasons, 42, 225, 232–234
Rhinoceros, 52, 297, 304
  beetle, 52, 53
Rib-faced deer, 23
Roaring, 8, 10, 104, 105, 255
Roe deer, 22, 23, 27–29, 44, 46, 47, 92,
    95–97, 103, 110, 113, 118, 123, 125,
    129, 134, 138, 146, 163, 194, 195,
    197, 198, 203, 204, 212, 249,
    251–253, 256, 258, 270, 276, 277,
    280, 285, 288, 289
Roe ring, 28, 103
Royal stag, 10
Ruminant, 6, 72
Rusa deer, 18, 19, 44
Rut, 7–10, 12, 17, 22–24, 26, 28, 29, 31,
    33, 44, 60, 63, 101–110, 112, 115,
    118, 163, 230, 238–240, 251, 255,
    257, 260, 271

S

Sable antelope, 55
Sambar deer, 13, 14, 44, 46, 201, 202
Scar, 3, 147, 148, 181, 184, 191
  hypertrophic, 280

Scent, 7, 102, 105, 110
Schomburgk's deer, 17, 18, 44
Scimitar-horned antelope, 55
Scoring system, 10, 37
Screwed on antler, 119, 200
Scrotum, 268, 269
Scur, 213
Season, 3, 17, 21, 44, 118, 127, 212,
    230–239, 244, 246
Selenium, 261
Seminiferous tubule, 250, 251, 269
Senile antlers, 197, 198
Sex chromatin, 288
Sexual dimorphism, 2, 101, 117, 285
Shedding
  of horn sheath, 62–65, 81
  premature, 253, 255
  of velvet, 9, 10, 12, 23, 28, 30, 32, 36,
    42, 44, 84, 102, 103, 108,
    111–113, 125, 130, 143, 144, 147,
    160, 161, 166, 167, 199, 200, 203,
    207, 231, 232, 236, 237, 239, 240,
    251, 259, 269, 278, 286–288, 293
Sheep
  Bighorn, 55, 59, 60, 65, 107
  Dorset, 55, 57
  four-horned, 213
Sika deer, 7, 10, 11–13, 44, 46, 115, 127,
    136, 141, 144, 148, 151, 154, 155,
    163, 194, 199–201, 202, 203, 205,
    207, 211, 231, 234, 236, 237,
    242–246, 251, 253, 255, 261, 273,
    285, 289–291, 298–300
  Formosan, 12
Sitka deer, 36
Sivatherium, 74, 109
Skin, 66–68, 123, 124, 127–130,
    142–145, 147–153, 173, 179, 181,
    182, 184, 186, 187, 189, 191, 207,
    213, 294, 300
Smell, sense of, 7, 105, 110
Social hierarchy, 102, 107, 108, 110, 116,
    285, 293
Social significance, 101–119, 285
Solstice, 232, 237, 238, 240, 242, 244,
    245
South America, 35, 38–42, 44, 93, 95, 96,
    241, 245, 246, 254
South Georgia Island, 225, 232
Sparring, 103, 104

Spaying, 127, 255, 278, 290–292
Spermatogenesis, 241, 243, 249, 250, 252, 269, 303
Spike buck, 260, 262
Spongiosa, 152, 154, 155, 159
Spotted deer, 298
Spotting, 7, 11, 20, 30, 31, 36, 115, 243
Springbok, 55
Squamous cell carcinoma, 215
Stephanocemos, 85, 86
Stress, 258, 303
^{90}Sr, 162
Superantler, 98, 208–210
Svalbard, 230, 257
Swamp deer, 14, 15, 44, 96
Syndyoceros, 73, 109

T

Taruca, 41
Teeth, 6, 23, 25–27, 52, 58, 69, 70, 72, 85, 105, 114, 117, 173, 219
  eruption, 69
Telemetacarpalia, 44–46, 95, 96
Temperature, 109, 110, 162, 163, 243, 269
Territoriality, 102–104, 110, 111, 117, 118, 186
Testis, 39, 203, 220, 231, 241, 243, 250–252, 259, 261, 268–270, 277–288, 302, 303
  atrophy, 241, 250, 251, 261, 269
  hypertrophy, 259
Testosterone, 3, 4, 60, 101, 104, 115, 119, 127, 139, 198, 200, 241, 242, 250–257, 259, 269–271, 274, 278–281, 288, 290–292, 302
Thailand, 18, 44
Thamin, 16, 17
Thermal
  adaptation, 243
  radiation, 109, 110, 163
  regulation, 162, 163
Threshing, 103, 111, 166, 222
Thyroid, 258
Thyroxine, 258, 261
Tine, 10, 16, 22, 34, 37, 74, 85, 89, 90, 97, 106, 107, 134, 135, 155, 160, 194, 204, 218, 219, 222–228
  bez, 10, 37
  trez, 10

Trabeculae
  bony, 67, 123, 124, 159, 160
  cartilaginous, 124, 153, 157, 158
Transequatorial shift, 10, 42, 225, 226, 232
Transplantation
  of antler bud, 146, 147
  of ear, 144–146
  of pedicle, 206
  of periosteum, 123, 129–131
  of skin, 66, 67, 129, 143, 144, 148, 184, 189, 191, 294
Triceratops, 52, 53
Tropics, 4, 13–21, 29, 38–40, 44, 60, 75, 76, 86, 109, 113, 115, 118, 230, 240, 241, 244–246, 252, 254, 272
Tuberosity, 272, 274, 275
Tufted deer, 24, 25, 44, 46
Tumor, 272–281, 288
Tundra, 33, 92, 222, 230, 300
Tunica media, 165, 166
Tusk, 6, 23, 25–27, 52, 69, 70, 72, 75, 76, 85, 105, 114, 117, 219, 297
Twin, 25–27, 31, 36, 220, 288, 289
Two antlers per year, 23, 28, 231, 234, 252

U

Unicorn, 66, 67, 70, 214, 215, 297, 298
Unilateral antler growth, 200, 202, 242, 243
Urine, 103, 105

V

Vascularization, 162–166, 280, 294
Vasectomy, 270
Velvet, 9, 10, 12, 23, 28, 30, 32, 36, 42, 44, 82, 84, 102, 103, 111, 112, 124, 131, 141, 143–145, 147–154, 164, 166, 203, 237, 279
Villafranchian, 87
Virgin Islands, 39
Vision, 7, 87, 222, 259, 260, 276
Vocalization, 8, 10, 13, 24, 29, 104

W

Wallow, 29, 31, 103, 111
Wapiti, 1, 3, 8, 10, 29–31, 44, 46, 47, 94, 95, 102–104, 106, 111, 118, 125,

Wapiti (*continued*)
134, 148, 150, 166, 210, 251, 259,
260, 272, 285, 288, 291, 293, 294
White-tailed deer, 35–39, 44, 95, 125,
127, 139, 152, 161, 167, 194–196,
208–210, 238–241, 254–257, 259,
261–263, 269, 274–276, 285, 286,
288–291, 293
Winter antler growth, 23, 28, 110, 113
Wisent, 65
Worm, 174, 221, 259
Wound healing, 3, 111, 112, 142, 144,
147, 148, 173, 174, 178, 179, 181,

184, 199, 201, 206, 207, 210, 272,
290, 294, 302

Y

Yearling, 125, 126, 237, 244, 249, 262,
293

Z

Zinc, 261
Zone of chondrification, 158
Zone of proliferation, 155, 156